THE STRUCTURE OF WESTERN EUROPE

by

J.G.C. ANDERSON

Professor of Geology, University College, Cardiff

PERGAMON PRESS

OXFORD · NEW YORK · TORONTO · SYDNEY · PARIS · FRANKFURT

U.K.	Pergamon Press Ltd., Headington Hill Hall, Oxford OX3 0BW, England
U.S.A.	Pergamon Press Inc., Maxwell House, Fairview Park, Elmsford, New York 10523, U.S.A.
CANADA	Pergamon of Canada Ltd., 75 The East Mall, Toronto, Ontario, Canada
AUSTRALIA	Pergamon Press (Aust.) Pty. Ltd., 19a Boundary Street, Rushcutters Bay, N.S.W. 2011, Australia
FRANCE	Pergamon Press SARL, 24 rue des Ecoles, 75240 Paris, Cedex 05, France
FEDERAL REPUBLIC OF GERMANY	Pergamon Press GmbH, 6242 Kronberg/Taunus, Pferdstrasse 1, Federal Republic of Germany

Copyright © 1978 J. G. C. Anderson

All Rights Reserved. No part of this publication may be reproduced, stored in a retrieval system or transmitted in any form or by any means: electronic, electrostatic, magnetic tape, mechanical, photocopying, recording or otherwise, without permission in writing from the publishers.

First edition 1978

British Library Cataloguing in Publication Data

Anderson, John Graham Comrie
The structure of Western Europe.—
(Pergamon international library).
1. Geology — Europe, Western
I. Title II. Series
554   QE260   77-30389
ISBN 0-08-022045-2 (Hardcover)
    0-08-022046-0 (Flexicover)

In order to make this volume available as economically and as rapidly as possible the author's typescript has been reproduced in its original form. This method unfortunately has its typographical limitations but it is hoped that they in no way distract the reader.

Printed in Great Britain by Cox & Wyman Ltd, Fakenham

PERGAMON INTERNATIONAL LIBRARY
of Science, Technology, Engineering and Social Studies
*The 1000-volume original paperback library in aid of education,
industrial training and the enjoyment of leisure*
Publisher: Robert Maxwell, M.C.

THE STRUCTURE OF WESTERN EUROPE

THE PERGAMON TEXTBOOK
INSPECTION COPY SERVICE

An inspection copy of any book published in the Pergamon International Library will gladly be sent to academic staff without obligation for their consideration for course adoption or recommendation. Copies may be retained for a period of 60 days from receipt and returned if not suitable. When a particular title is adopted or recommended for adoption for class use and the recommendation results in a sale of 12 or more copies, the inspection copy may be retained with our compliments. The Publishers will be pleased to receive suggestions for revised editions and new titles to be published in this important International Library.

Other titles of interest:

ALLUM
Photogeology and Regional Mapping

ANDERSON & OWEN
The Structure of the British Isles

BOWEN
Quaternary Geology

BROWN et al
The Geological Evolution of Australia and New Zealand

CONDIE
Plate Tectonics and Crustal Evolution

OWEN
The Geological Evolution of the British Isles

PRICE
Fault and Joint Development in Brittle and Semi-Brittle Rock

SIMPSON
Geological Maps

SPRY
Metamorphic Textures

CONTENTS

Preface ... ix

Acknowledgements ... x

Abbreviations used in References ... xi

Chapter 1. INTRODUCTION
1.1 Historical Background ... 1
1.2 Outline of the Geology of Western Europe ... 1
1.3 The Stratigraphical Sequence in Western Europe ... 3
1.4 The Structural Sequence in Western Europe ... 3

Chapter 2. PRECAMBRIAN BLOCKS
2.1 The Baltic Shield (Sweden, parts of Norway and Bornholm, Denmark) ... 8
2.2 Southern Norway ... 14
2.3 North-West Scotland ... 20

Chapter 3. CALEDONIAN FOLD-BELTS (MAINLY METAMORPHIC)
3.1 Introduction ... 28
3.2 The North-West Caledonian Front ... 28
3.3 The Northern Highlands and the Orkney Islands ... 34
3.4 The Grampian Highlands (Scotland and Ireland) ... 40
3.5 Connemara and its Flanks (Ireland) ... 50
3.6 Shetland ... 54
3.7 The Caledonian Front in Scandinavia (Norway and Sweden) ... 56
3.8 The Caledonian Fold-Belt in Scandinavia (Norway and Sweden) ... 61

Chapter 4. CALEDONIAN FOLD-BELTS (MAINLY SEDIMENTARY)
4.1 Introduction ... 73
4.2 Wales and the Welsh Borderland ... 73
4.3 The Lake District and the Isle of Man ... 81
4.4 The Southern Uplands in Scotland and Ireland ... 86
4.5 South-East Ireland ... 91
4.6 The Brabant Massif (Belgium) ... 94

Chapter 5. HERCYNIAN FOLD-BELTS (LARGELY METAMORPHIC)
5.1 The Massif Central and the Montagne Noire - Introduction ... 98
5.2 The Massif Central ... 99
5.3 La Montagne Noire ... 101
5.4 The Variscan Arc - Introduction ... 101
5.5 The Vosges and the Schwarzwald ... 103
5.6 The Rheinisches Schiefergebirge and the Ruhr Coalfield ... 105
5.7 The Armorican Arc - Introduction ... 105

Contents

5.8	The Armorican Arc - France and the Channel Islands	107
5.9	The Armorican Arc - South-West England	108
5.10	The Armorican Arc - Southern Ireland	111
5.11	The Ardennes and the Belgian Coalfield	112
5.12	The Iberian Peninsula - Introduction	115
5.13	The Cantabrian and León-West Asturias Zones	115
5.14	The Central Iberian and the Ossa-Morena and South Portuguese Zones	118

Chapter 6. HERCYNIAN SEDIMENTARY BASINS, BLOCKS AND COALFIELDS

6.1	Introduction	123
6.2	The Midland Valley of Scotland	123
6.3	North-East England	128
6.4	Central England	132
6.5	South Wales and Pembrokeshire Coalfields and their Borders	137
6.6	The Bristol and Forest of Dean Coalfields and their Margins	142
6.7	Central Ireland	144

Chapter 7. ALPINE FOLD-BELTS

7.1	Introduction - The Western Alps	151
7.2	The Jura Mountains	155
7.3	The Franco-Swiss Plain or Molasse Basin	157
7.4	The Swiss-Italian Alps - The Pre-Alps	157
7.5	The Swiss-Italian Alps - The Helvetides and Ultrahelvetides	160
7.6	The Swiss-Italian Alps - The Hercynian or External Massifs	163
7.7	The Swiss-Italian Alps - The Pennides	163
7.8	The Northern Franco-Italian Alps - The Helvetides or Subalpine Chains	166
7.9	The Northern Franco-Italian Alps - The Hercynian or External Massifs and the Pennides	166
7.10	The Central Franco-Italian Alps - The Dauphinois Zone	168
7.11	The Central Franco-Italian Alps - The Hercynian or External Massifs	170
7.12	The Central Franco-Italian Alps - The Pennides	172
7.13	The Southern French Alps - The Provençal Chains	174
7.14	The Southern French Alps - The Hercynian or External Massifs	177
7.15	The Pyrenees (France and Spain) - Introduction	177
7.16	The Pyrenees - Northern External and Satellite Massifs Zones	180
7.17	The Pyrenees - Axial Zone	182
7.18	The Pyrenees - Southern External (Nogueras) and Sub-Pyrenaic Sierras Zones	183
7.19	The Pyrenees - The Biscayan Zone	185
7.20	The Catalonian and Iberian Chains (Spain)	185
7.21	The Betic Cordillera - Introduction	187
7.22	The Prebetic and Subbetic Zones	189

Contents

	7.23	The Betic Zone, The Gibraltar Complex and the Straits of Gibraltar	189

Chapter 8. MESOZOIC AND TERTIARY SEDIMENTARY BASINS OIL/GAS FIELDS

8.1	The Saône Graben and the Rhône Corridor (France)	193
8.2	The Rhine Rift Valleys (France and West Germany)	195
8.3	The Ebro and Guadalquivir Basins (Spain)	199
8.4	The Aquitaine and Paris Basins and the Low Countries (France, Belgium, Holland, Denmark and West Germany) - Introduction	200
8.5	The Aquitaine Basin	203
8.6	The Paris Basin	204
8.7	The Low Countries	208
8.8	Eastern and Southern England	210
8.9	Basins and Grabens of the North-West Continental Shelf - Introduction	216
8.10	The North Sea	216
8.11	The English Channel, the Western Approaches and the Sea-Floor off North and North-West Scotland	218

Chapter 9. TERTIARY AND QUATERNARY VOLCANIC STRUCTURES

9.1	Introduction	222
9.2	The British Isles	222
9.3	The Massif Central of France	227
9.4	The Rhineland (West Germany)	231

Chapter 10. SUMMARY OF THE STRUCTURES OF WESTERN EUROPE AND THEIR SIGNIFICANCE IN PLATE TECTONICS MODELS

10.1	Distribution and Inter-relationship of Western European Structures	235
10.2	Plate Tectonics Models	238

Index 244

ANDERSON
STRUCTURE OF WESTERN EUROPE
ERRATA SLIP:
on page 19 the captions for the
2 photographs have been transposed

PREFACE

Almost the whole geological succession and nearly all the structural events which have affected the crust during and since Archean times are represented in Western Europe. This was one of the reasons why geology was born in Western Europe and why geological principles were established which have influenced geological discovery throughout the world.

The rocks and structures, moreover, are spectacularly displayed in landscapes varying from the rocky Atlantic coastlines to the mountain cliffs of Norway, the British Isles, the Alps, the Pyrenees and the Betic Ranges.

Geological research carried out for 200 years is extensively recorded in all Western European languages, but there are few accounts in English, or any other language, of the geological structure as a whole, designed for the needs of students and teachers of geology in schools and universities and to provide guidance for geologists from other parts of the world.

As far as possible, within its wider terms of reference, the book follows the plan successfully adopted in the Pergamon publication 'The Structure of the British Isles' by Professor J.G.C. Anderson and Professor T.R. Owen, to which the reader is referred for details regarding these islands which are not included in this book because of space considerations. The question of space, and therefore of cost, has, in fact, led in general to the cutting out of some details which might otherwise be included; numerous references are, however, provided.

There is no firm political or geographical definition of Western Europe and in deciding on the scope of the book in this respect consideration had again to be given to length. The following countries are included: Norway, Sweden, Denmark, Holland, United Kingdom, Eire, Belgium, Luxembourg, France, Western Germany, Switzerland, Spain and Portugal. Where necessary, references are made to the geology of adjacent countries and, particularly for the Alps, to parts of Austria and Northern Italy.

It is clearly impossible in a book of this kind to describe local outcrops, but to help readers who wish to study the geology in the field descriptions are given of the topography in each district and of the degree of rock exposure; references are made to excursion guides in several languages.

The geological structure of parts of the sea-floor off Western Europe, the sources of rich oil and gas deposits, are also described.

A final chapter summarises the geological structure and evolution in the light of Plate Tectonics.

Metric units are used throughout. However, for the British Isles, where Imperial units appear on nearly all published maps, memoirs, etc., both Imperial units and metric units are given.

ACKNOWLEDGEMENTS

The author wishes to express his gratitude to numerous colleagues for many helpful suggestions and stimulating discussions. Particular thanks are due to Professor T. R. Owen, University College of Swansea, and Dr. R. A. Gayer, University College of Cardiff.

Thanks are also due to Miss Liza Morgan for typing the manuscript and to Mr. W. S. Thomas and Mr. A. Dean who drew most of the illustrations.

ABBREVIATIONS USED IN REFERENCES

Adv. Sci.	Advancement of Science
Amer. J. Sci.	American Journal of Science
Am. Soc. Pet. Geol.	American Association of Petroleum Geologists
Ann. Soc. Géol. Belgique	Annales de la Société Géologique de Belgique
Bol. Inst. Geol. Minero Espan.	Boletin del Instituto geologico y minero de España
Bull. Geol. Inst. Uppsala	Bulletin of the Geological Institution of the University of Uppsala
Comp. Rend. de Congrès Intern. Strat. Géol. Carbonif.	Comptes rendus du Congrès international de Stratigraphie et de géologie du Carbonifère
Consejos Super. Cient. Inst. Lueas	Consejos Superior Cientificas Instituto Lueas
Cumb. Geol. Soc.	Cumberland Geological Society
Earth Planet. Sci. Lett.	Earth and Panetary Science Letters
Earth-Sci. Rev.	Earth-Science Review
Eclogae Geol. Helv.	Eclogae Geologicae Helvetiae
Econ. Geol. Mono.	Economic Geology Monographs
Edin. Geol. Soc.	Edinburgh Geological Society
Eiszeitalter Gegenw.	Eiszeitalter und Gegenwart
Geol. Fören. Stockholm Förh.	Geologiska föreningens i Stockholm förhandlingar
Geol. Rundschau	Geologische Rundschau
Geol. Soc.	Geological Society
G.M.	Geological Magazine
Guides Géol. Rég.	Guides Géologiques Régionaux
I.G.S. Report	Institute of Geological Sciences Report
Journ. Geol.	Journal of Geology
Journ. Geol. Soc. (J.G.S.)	Journal of the Geological Society (London)
Leid. geol. Meded.	Leidsche geologische Mededelingen
Liverpool and Manch. Geol. J.	Liverpool and Manchester Geological Journal
Matér. Carte. Géol. Suisse	Matériaux cartographique pour la géologie de la Suisse
Mém. Ac. des Sc.	Mémoires de l'Academie des Sciences (Paris)
Mém. Bur. Réch. Géol. Minières	Mémoires du Bureau de Récherches Géologiques et Minières
Mém. Soc. vaud. Sci. Nat.	Mémoires de la Société vaudoise des sciences naturelles (Lausanne)
M.G.S.	Memoirs of the Geological Survey of Great Britain
Nat.	Nature (London)
Nat. (Phys. Sci.)	Nature (Physical Sciences)
Neues Jahrbuch Mineral. Abhandl.	Neues Jahrbuch für Mineralogie. Abhandlungen
P.G.A.	Proceedings of the Geologists' Association

P.R.I.A.	Proceedings of the Royal Irish Academy
Proc. Geol. Soc.	Proceedings of the Geological Society (London)
Proc. York. Geol. Soc.	Proceedings of the Yorkshire Geological Society
Q.J.G.S.	Quarterly Journal of the Geological Society (London)
Réun. Ann. des Sci. de la Terre	Réunion Annuel des Sciences de la Terre
Rev. des Sci. Nat. d'Auvergne	Revue des Sciences Naturelles d'Auvergne
Sci. Proc. Roy. Dublin Soc.	Scientific Proceedings of the Royal Dublin Society
Sci. Prog.	Science Progress
Scot. Journ. Geol.	Scottish Journal of Geology
Sveriges geol. Unders.	Sveriges Geologiska Undersoekning
T.G.S.G.	Transactions of the Geological Society of Glasgow
Trans. Leicester Lit. & Phil. Soc.	Transactions of the Leicester Literary and Philosophical Society
T.R.S.E.	Transactions of the Royal Society of Edinburgh
Verhandl. Naturforsch. Ges. Basel	Verhandlungen der Naturforschenden Gesellschaft in Basel

Chapter 1
INTRODUCTION

1.1   Historical Background

Geology was conceived and born in Western Europe, and for about 200 years European geologists have established principles which have influenced geological discovery throughout the world. Hutton (1795), in Edinburgh, wrote his Theory of the Earth, generally regarded as the foundations of modern geology. Werner, about the same time, in Germany, in spite of his theoretical errors, established the principles of mineralogy. Guettard, in France, produced what amounted to the first geological map in 1751, and demonstrated that extinct volcanoes could be recognised by their rocks and forms. William Smith's Geological Map of England (1815) was another of the great stepping stones in geological science.

Later in the nineteenth century, the recognition of the great overthrusts of N.W. Scotland and of the nappes of the Alps provided the foundations of modern tectonics. Similar structures were recognised at an early date in the Caledonides of Scandinavia, and in the Baltic Shield of the same region light was thrown on the complex processes of high-grade (catazonal) metamorphism, migmatisation and granitisation.

The geological principles and methods established in Europe were very soon applied to other continents, leading to further fundamental discoveries which provided a feed-back and stimulus to European geologists themselves.

The foundation of geology in Western Europe, about the end of the eighteenth century, was not due to chance. It arose partly from the advanced state, for the times, of European technology, including mining and civil engineering, and partly to the fact that men with questing minds, trained in technology, began to take an interest in the remarkably varied formations and structures of Western Europe, wider ranging than those of any comparable area of the world.

1.2   Outline of the Geology of Western Europe

Almost every subdivision of the stratigraphical succession, with the possible exception of the earliest Precambrian, is present in Western Europe, and many occur in contrasting marine and sedimentary facies. No active volcanoes occur in Western Europe as defined (see Preface), but are present in Italy. Moreover, the last of the Quaternary volcanoes of Central France (9.3) became extinct only 3000 years ago or even less.

Consequently, nearly all the major stratigraphical divisions are named after a European locality (e.g. Jurassic) or after a European facies (e.g. Cretaceous).

Nearly all the structural events which have affected the crust since Archean times can be studied in Western Europe. In Scandinavia five major Precambrian orogenic cycles can be recognised and in N.W. Scotland three;

TABLE I

LATE PRECAMBRIAN (U. PROTEROZOIC) SUCCESSIONS AND POSSIBLE CORRELATIONS IN WESTERN EUROPE

Scottish Kratogen and Parautochthonous Nappes	Scandinavian Kratogen and Parautochthonous Nappes	Metamorphic Caledonian Fold-Belt of Scotland and Ireland	English Midlands Kratogen and its W. margin	Caledonian Fold-Belt of Wales	Hercynian Fold-Belt of France and Channel Islands
(*)		(†)	(*)	(†)	(*)
U. Torridonian	Eocambrian or Sparagmitian	M. Dalradian	Charnian	Arvonian	Brioverian
L. Torridonian		L. Dalradian	Longmyndian	Monian	
		Moinian	Uriconian		

(*) structural break below L. Palaeozoic
(††) passage into L. Cambrian in most districts
(†) passage into L. Cambrian

both these Precambrian Blocks are part of the "Eo-Europa" of some authors (e.g. Ager, 1975). In large regions of Norway and Sweden, and of the British Isles ("Palaeo-Europa"), Caledonian structures of several dates are displayed. In the central and southern British Isles, in several large massifs in France and Western Germany and in much of the Iberian Peninsula ("Meso-Europa") Hercynian structures predominate.

Intra-Mesozoic folds and faults, not as powerful in Europe as in some parts of the world, referred to in N.W. Germany and Holland as Saxonian, are features of the geology of these regions and of parts of E. and S. England.

Finally, in the Alps, the Pyrenees and in the Betic Cordillera ("Neo-Europa") the various phases of the Alpine orogeny have produced some of the most spectacular structures and scenery in the world.

1.3 The Stratigraphical Sequence in Western Europe

Although metasediments ranging back nearly 3000 m.y. are present in Western Europe, an understanding of Archean and Lower Proterozoic history mainly depends on working out the relationships of structural and metamorphic cycles and on radiometric dating.

Only in the Upper Proterozoic has it been possible to establish subdivisions on a valid stratigraphical basis. As these formations contain, at most, only sparse and poorly-preserved fossils there are no stratigraphical terms, analogous to those of the Phanerozoic, which are acceptable for the whole of Western Europe. Successions for some regions are given in Table I; inter-regional correlations are mostly tentative, apart from the youngest Precambrian where a widespread glacogene forms a marker horizon.

The most widely used subdivisions of the Phanerozoic are given in Table II. Where other stratigraphical names are used in regional descriptions, the terms shown on Table II are frequently added in brackets.

For the Lower Palaeozoic the classic British Series names are given. For the rest of the succession, however, many of the stage names are derived from the European mainland. In most of the British Isles the Carboniferous System is commonly divided into the Carboniferous Limestone (Avonian), Millstone Grit and Coal Measures, roughly equivalent to the Dinantian, Namurian and Westphalian. The long-established Old Red Sandstone and New Red Sandstone are used for the appropriate facies, particularly where it is not easy to separate the continental Permian and Triassic red beds (Rotliegendes).

In some regions the Palaeocene has not been differentiated from the top of the Cretaceous or the base of the Eocene.

Pleistocene deposits (largely, but not entirely, glacial in Western Europe) and Holocene deposits are not described, nor are the effects of glacial erosion, except insofar as they affect the display of structures. The structural implications of Quaternary vulcanicity are, however, dealt with in Chapter 9.

1.4 The Structural Sequence in Western Europe

The structural evolution of the Eurasian continental block began well over

TABLE II

PHANEROZOIC STRATIGRAPHICAL AND STRUCTURAL SEQUENCES IN WESTERN EUROPE

SYSTEM	"SERIES"	"STAGE"	AGE (m.y.)	MAIN TECTONIC EVENTS	
Neogene	Pliocene	Pontian	12	X	
	Miocene	Tortonian Helvetian	19		
		Burdigalian Aquitanian	26		
Palaeogene	Oligocene	Chattian Stampian Sanndisian	37	X	LATE ALPINE
	Eocene	Ludian Marinesian Auversian	43	X (Provencal)	
	Palaeocene	Lutetian Ypresian Sparnacian Thanetian Montian Danian	65	X	
Cretaceous	Upper Cretaceous	Maestrichtian Campanian Santonian Coniacian		X	EARLY ALPINE
		Turonian			
	Middle Cretaceous	Cenomanian Albian		X	
	Lower Cretaceous	Aptian Barremian Hauterivian Valangian Berriasian	135		
Jurassic	Upper Jurassic (Malm)	Purbeckian Portlandian Kimmeridgian Oxfordian Callovian		Young Cimmerian	SAXONIAN
	Middle Jurassic (Dogger)	Bathonian Bajocian Aalenian			
	Lower Jurassic (Lias)	Toarcian Pliersbachian Sinemurian Hettangian Rhaetian	195		

Senonian spans Maestrichtian–Coniacian. Neocomian spans Aptian–Berriasian.

TABLE II (continued)

PHANEROZOIC STRATIGRAPHICAL AND STRUCTURAL SEQUENCES IN WESTERN EUROPE

SYSTEM	"SERIES"	"STAGE"	AGE (m.y.)	MAIN TECTONIC EVENTS	
Triassic	U. Triassic (Keuper)	Norian Carnian		Old Cimmerian	
	M. Triassic (Muschelkalk)	Ladisian Anisian			
	L. Triassic	Werfenian	225	Palatine	
Permian		Thuringian Saxonian Autunian	280	Saalian	H E R C Y N I A N
Carboniferous	Upper Carboniferous	Stephanian Westphalian (s.s.) Namurian		Asturian	
	Lower Carboniferous (Dinantian)	Visean Tournaisian Strunian	345	Sudetic Bretonnic	
Devonian (Old Red Sandstone)	Upper Devonian	Famenian Frasnian			
	Middle Devonian	Givetian Eifelian		Svalbardian	
	Lower Devonian	Emsian Siegenian Gedinian	410	Brabancon	C A L E D O N I A N
Silurian		Ludlovian Wenlockian Llandoverian	440	Taconic	
Ordovician		Ashgillian Caradocian Llandeilian Llanvirnian Arenigian Tremadocian	530	X Sardic	
Cambrian	U. Cambrian M. Cambrian L. Cambrian		600	Cadomian	

3000 m.y. ago, but in Western Europe there is, as yet, no definite evidence of events older than about 2800 m.y. Nevertheless, in parts of the Baltic Shield outside of the region, notably in the Kola Peninsula in the U.S.S.R., Katarchean dates have been obtained (2.1). Metamorphism dated at about 2800 m.y. is the earliest of a number of recognisable metamorphic and structural events, mostly due to major orogenies, which took place in Western Europe during the 2000 m.y. or so of Archean and Lower Proterozoic times. Relative ages have been established by structural studies, and radiometric dates have been obtained in Scandinavia, Scotland and within the Hercynides in N.W. France and in the Channel Isles.

The dates of some of the events in different regions do not correlate; this especially applies to those in Scotland as opposed to Scandinavia. This is to be expected as it would seem that separation of these two parts of the European Precambrian went back well into Precambrian times. However, the coincidence of Pregothian dates (2.1) in Scandinavia, of Scourian dates (2.3) in Scotland and Icartian dates (5.10) in N.W. France and the Channel Islands suggest a very widespread cycle of events round about 2600 m.y. ago.

In the kratogen of N.W. Scotland (2.1) the Scourian cycle was followed by the Inverian cycle dated at about 2200 m.y. and the Laxfordian phases at from 1500 to 1600 m.y. After that there is no record of events until the deposition of the Lower Torridonian just under 1000 m.y. ago.

These dates do not show any close coinciding with Scandinavian events (2.1) following the Pregothian including the last major metamorphism in that region, the Sveconorwegian about 1000 m.y. ago. This, however, possibly correlates with the 1000 to 1200 m.y. old Penteverian metamorphism in Brittany (5.10).

In the Moinian of the Caledonian Fold-Belt of Northern Scotland, not far E. of the Border Thrust-Zone, pegmatites have been dated at 740 m.y. It is debatable, however, whether these are evidence of a significant structural event (for discussion see 3.4).

About the beginning of the Cambrian, Cadomian folding and metamorphism occurred in several parts of Western Europe, particularly in N.W. France (Cadomus = the modern Caen). Although in many parts of Western Europe there is no sedimentary or structural break at the base of the Cambrian, in other regions, including much of Britain, there is a discordance which is probably due to Cadomian movements.

Cambrian times, for the most part, were tectonically relatively quiet, but at the end of the period Cambrian and earlier sediments were affected by Sardic events which can be regarded as an early phase of the Caledonian orogeny. Just how widespread these were in the Western European Caledonides is debatable (3.4).

Some of the most powerful and widespread Caledonian movements and metamorphism occurred in the Ordovician, particularly in the middle of the period and at the end - Taconic (from the Taconic Mountains, eastern U.S.A.). The earlier Caledonian phases in the northern British Isles have been referred to as the Grampian Orogeny. During most of the Silurian tectonism was relatively subdued, but towards or at the end of the period further folding, accompanied by at least low-grade metamorphism, was widespread in the Caledonides. These movements are sometimes referred to as Brabancon as the main Caledonian folding took place in the Brabant Massif of Belgium at that time. However, powerful folding on Caledonian trend-lines recurred, particularly in Scotland and Norway, after the deposition of the Lower Old Red Sandstone. These Svalbardian (Svalbard = Spitsbergen) events took place some 370 m.y. ago.

In the widest sense, therefore, Caledonian tectonism took place for some 160 m.y. The fact that the final Svalbardian events are separated by only some 20 m.y. from the Bretonnic phase (early Hercynian) bears witness to the frequent instability of the Western European crust.

These Bretonnic movements took place some 340 to 345 m.y. ago in Brittany and some other regions but do not seem to have affected many parts of Western Europe where the Carboniferous often follows the Devonian without break. The most powerful and widespread Hercynian structural events, accompanied in some regions by metamorphism, took place about the end of the Dinantian, some 325 m.y. ago, and at the end of the Stephanian, some 295 m.y. ago. These are termed respectively the Sudetic (from the Sudetes mountains) and the Asturian (from Northern Spain).

The Saalian (from Saal in S. Germany) tectonism, which took place in the Permian some 260 m.y. ago, often makes it possible to separate the Autunian stage, in which coal formation in some parts of Western Europe took place, from the Saxonian stage with its continental red beds. If the more localised Palatine (from the German Rhineland) phase (225 m.y.), at the end of or immediately after the Permian, is regarded as the last of the Hercynian orogeny, the latter, in the wide sense, has a time-span of about 120 m.y.

Middle Triassic and Jurassic movements in the Alps and in other parts of Western Europe, including intra-Jurassic movements such as the Saxonian (from Saxony, N.W. Germany) or Cimmerian disturbances, can all be regarded as precursors of the more powerful Alpine tectonism of late Mesozoic and Tertiary times. These movements occurred, or were at their peak, at different times in various parts of the Alps, but broadly fall into the Early Alpine phases, from the middle Cretaceous to the early Palaeocene, and the Late Alpine phases, from the Eocene to the beginning of the Pliocene. The Early phases were responsible for the structures of the Internal Alpine Zone and the Late phases for those of the External Zone, with the Eocene movements predominating in Provence.

The late movements, in a milder form, also produced the basins of the Western European seaboard, of the North Sea and of Southern England. As late as the Pontian, about 10 m.y. ago, thrusting occurred on the W. margin of the Jura.

Differential uplift and depression continued at least until the end of the Pliocene (3 m.y.). In fact, vertical movements, mainly isostatic and connected with glaciations, went on throughout the Pleistocene, and in northern Scandinavia are probably still going on.

REFERENCES

Ager, D.V. 1975. The geological evolution of Europe. P.G.A., 86, 127-154.
Hutton, J. 1795. <u>Theory of the Earth with Proofs and Illustrations</u>.
   2 vols. Creech, Edinburgh.

Chapter 2
PRECAMBRIAN BLOCKS

2.1 The Baltic Shield - Sweden, parts of Norway and Bornholm (Denmark) (Fig. 2.1)

The Baltic or Fennoscandian Shield is the largest outcrop of the Precambrian "basement" in Europe. It also forms the frame or kratogen on the E. and S.E. margin of the N.W. European Caledonides (3.1). Practically the whole of Sweden and parts of N.E. and S.E. Norway belong to the Shield, as well as Finland and nearby districts of Russia.

The W. boundary of the region described in the present section is defined by the Caledonian front from the Barents Sea to near Hamar in Norway and from there to the Skagerrak by the E. border of the Oslo Graben; the Danish island of Bornholm is also included. Southern Norway is described below (2.2).

Much of the region, especially around the Baltic and the Gulf of Bothnia, is a low peneplain modified by Pleistocene glaciation. Excellent exposures occur along the coasts and in innumerable roches moutonnées and glacially-smoothed surfaces. There are vast areas of coniferous forest (taiga) which beyond the Arctic Circle merge into birch scrub and then tundra. In the S. of Sweden the glacial drift becomes more continuous and there is also some sedimentary cover. It is only in the W., towards the Caledonian front, that the Shield rocks reach any great height.

Excursions are described in the following guides of the International Geological Congress: A21 (Mesozoic, S. Sweden); A22 (Silurian, Gotland); A23 (L. Palaeozoic, Central Sweden); A24 (thrust region, S. Sweden).

The geological succession is as follows :-

	Palaeocene
	Cretaceous
	Jurassic
	Triassic
	Silurian
	Ordovician
	Cambrian
Upper Proterozoic	(Eocambrian or Sparagmitian
	(Visingso formation
	((Dalslandian
	((Jotnian
Lower Proterozoic	(
	((Sub-Jotnian
	((Gothian
	(
	(Svecofennian (Svionian)
Archean	Pregothian

Fig. 2.1 GEOLOGICAL MAP OF BALTIC SHIELD AND ITS MARGINS
B.F. = Bamble formation;   D   = Dalslandian;
K.F. = Kongsberg formation; O.G. = Oslo Graben;
R.   = Rogaland;   S   = Scania

The last major metamorphism, the Sveconorwegian, widespread in the southern part of the region, took place about 1000 m.y. ago.

Pre-Eocambrian

The older Precambrian subdivisions, as in other parts of the world, are not stratigraphical but are complexes formed during prolonged cycles of sedimentary, tectonic, metamorphic, migmatic and intrusive events. Within each there may be several sub-cycles. Moreover, the events of a later cycle are often superimposed on those of an earlier and, in fact, the rocks of a younger complex may have been regenerated from those of an older.

The broad distribution and age-ranges of the complexes are as follows :-

Dalslandian	Southern Swedish-Norwegian border	1000 m.y.
Jotnian	Central Sweden and southern Swedish-Norwegian border	1200 - 1300 m.y.
Sub-Jotnian	Central Sweden and southern Swedish-Norwegian border	1670 m.y.
Gothian	S. Sweden, S.E. Norway and Bornholm	1300 - 1750 m.y.
Karelian	N. Sweden and N. Norway	1750 - ?2000 m.y.
Svecofennian	Central and N. Sweden	
Pregothian	S.W. Sweden, N. Sweden and N. Norway	2500 m.y.

Katarchean dates have been obtained in parts of the Baltic Shield, notably the Kola Peninsula in the U.S.S.R., but no rocks of this age have been identified in the Swedish/Norwegian section of the Shield. The oldest rocks are those of S.W. Sweden, referred to as Pregothian and considered to be equivalent to the Prekarelian of Finland; some rocks of equivalent age also occur in the far north (see below).

The Pregothian rocks are mainly gneisses, predominantly granitic but including more basic members now largely amphibolites; pegmatites are also present. The gneisses seem to have been intruded as plutonic rocks in the sediments, including quartzites and calcareous metasediments. About 70 km S. of Gothenberg charnockite occurs. North of Lake Vänern there are wide dykes and sheets of altered olivine-gabbro known as hyperite. In one of these bodies, at Taberg, there is a titaniferous iron-ore deposit. The general strike of the gneissose structures is northerly, and along the E. margin of the Pregothian there is a schistose belt dated at from 1200-1300 m.y. North of Lake Vänern the Pregothian gneisses are themselves cut by a northerly trending belt of highly schistose or mylonitic rocks.

The Svecofennian extends from Central Sweden to Norbotten in the far north. The detailed studies which have been carried out in Central Sweden have revealed four main stages of the cycle: deposition - on an unknown basement - of the Leptite formation of volcanics and sediments; folding, and the intrusion of the early Svecofennian granites; basic dyke intrusion in a non-orogenic phase; migmatisation and intrusion of the late Svecofennian granites.

The Leptite formation consists of metamorphosed fine-grained sodic and potassic volcanics; the very fine-grained rocks are known as halleflintas and the relatively coarser as leptites. Limestones and dolomites also occur, as well as beds of sedimentary iron-ore often of the quartz-banded type.

The Leptite formation has undergone open folding in the western part of the region but isoclinal folding further east. The frequent absence of bedding adds to the difficulties of structural interpretation; lineation studies have, however, shown that there have been at least two fold-phases.

The first folding was closely connected with the intrusion of the early Svecofennian granites; some basic rocks, now mostly uralite gabbros, are also found.

This intrusive phase was followed by a non-orogenic, relatively cool interval during which long, straight basic dykes were intruded. These are clearly earlier than the last stage of the Svecofennian cycle. This was one of granite intrusion and migmatisation which affected large parts of all the older formations - the Svecofennian regeneration as it is sometimes termed. Small scale plastic folding is common but the apparent absence of general folding led Magnusson (1940) to state that the Late Svecofennian was a time of downwarping rather than orogeny. The Svecofennian continues without much variation into the S. of Norrland but changes as it is traced further N. into Vasterbotten. In the northern part of this county, in the Skelleft Field, there is an extensive development of low-grade metasediments and metavolcanics, folded with a general W.N.W. strike. The metasediments include basaltic to rhyolitic volcanics along with pelites, psammites, some limestone beds and conglomerates. Important sulphide deposits, probably connected with migmatisation in the late Svecofennian and occurring in an anticlinal zone, include pyrites, chalcopyrites, zincblende, mispickel and subordinate galena; gold and some silver are also mined.

Considerable attention has been given to the Precambrian rocks of Norbotten in the extreme N., partly because of the very important iron deposits. Here the northerly-striking Svecofennides meet the Karelides which extend N.W. from Karelia in the U.S.S.R. across Finland to Arctic Sweden and Norway. Formerly the Karelian cycle was regarded as younger than the Svecofennian, but it is now thought that they are roughly contemporaneous although the Karelian involves different structures and a different sedimentary facies. The Svecofennides contain more pelites and psammites while the Karelides contain more pure quartz sandstones and carbonates; basic metavolcanics occur in both. The Svecofennides could thus represent the eugeosyncline, and the Karelides the miogeosyncline, of one cycle.

Traced N., in fact, into Norbotten the Svecofennides show some of the characteristics of the Karelides, and Padget (1973), in a summary of the Precambrian of Northern Sweden, refers to Svecofennian rocks as Karelian type. Seven main stages are distinguished in the geological evolution of the region. North of Kiruna there is a cratonic nucleus of northerly-trending gneisses dated at 2700 m.y., i.e. comparable to the Pregothian. On this was deposited a Svecofennian sedimentary/volcanic sequence in which the preservation of "way-up" structures has made it possible to determine the order of deposition.

Thirdly came deformation and the intrusion of the Haparanda granite suite dated at 1880 m.y. Acid volcanics-porphyries - (1650 - 1600 m.y.) were followed by further deformation and also block faulting. The sixth phase was the formation of potassic granites and related rocks (about 1550 m.y.) which occupy some 50% of the region; these show both migmatic and discordant relationships to older rocks. Lastly there was faulting, shearing and block movements.

The huge Kiruna iron-ore deposit, worked since 1903, was intruded magmatically as a sill between two of the porphyry flows mentioned above. The ore consists of magnetite with a variable amount of flourapatite (Park and MacDiarmid, 1970, 246).

The Karelides extend W. into E. Finnmark, in N. Norway (Bugge, 1960, 78-92). Here a Prekarelian basement, seen S. of the Varangerfjord, of amphibolite, mica-gneiss, granite-gneiss and granite is also present. The Karelian metasedimentary/metavolcanic succession starts with a basal conglomerate and includes quartz-banded iron-ores worked in the Bjørnevann open-pit S. of Kirkenes.

A first phase of folding with overturning towards the S.W. and pitching towards the S.E. was followed by cross-folding with shearing and easterly pitch.

On the eroded surface of these older Karelian rocks late or post-Karelian sediments and extrusives were deposited - the Petsamontunturit formation; ultrabasic and basic intrusion took place. Large deposits of nickeliferous pyrrhotite ores occur, worked at Petsamo in the U.S.S.R.

Rocks ascribed to the Gothian cycle occupy a large part of S.E. Sweden. Here the Småland granites are intruded into acid volcanics and quartzitic sediments. These all lie E. of the Pregothian rocks, S. of Lake Vänern, described above. To the W. of these older rocks the Åmål Complex, extending N. of Gothenberg with dominant northerly strike, is placed in the Gothian. This is made up mostly of granite intruded into metasediments and metavolcanics deposited on a Pregothian gneiss basement. Metamorphism of the Åmål formation left its mark on this basement (Geijer, 1963, 109). Further W. still, the Stora Le-Marstrand series of often highly metamorphosed quartzites, pelites and metavolcanics is regarded as Gothian although older than the Åmål formation.

The part of Norway E. of the Oslo Fjord is also largely made up of Gothian rocks.

Outcropping mainly near the S. part of the Swedish-Norwegian border, three supracrustal formations - the Dalslandian, sub-Jotnian and Jotnian - have been considered to occur stratigraphically in the order just given and to be all younger than the Gothian.

The sub-Jotnian includes the Dala porphyries, agglomeratic tuffs and rapakivi-type granites (more extensively developed in Finland). Strong denudation preceeded the deposition of the Jotnian sediments, largely red clastics, including, along with shales, the Dala sandstone of S.W. Sweden and the corresponding sandstone of the Trysil district (Dons, 1960, 58) of S.E.

Norway, about 100 km N.E. of Lake Mjøsen; diabase-lavas are also present. The Jotnian rocks are mostly flat in Sweden but have easterly dips in Norway with increasing deformation westwards, i.e. on approaching the Caledonian front.

The Dala porphyries have been dated at 1670 m.y. (Welin and Lundqvist, 1970) and the Dala sandstone at 1200 to 1300 m.y. (Magnusson, 1965, 6). The sub-Jotnian, at any rate, must therefore be part of the Gothian cycle and older than the Dalslandian (see discussion by Geijer, 1963, 125).

The Dalslandian formation consists of an older series (Kappebo) made up of conglomerate, greywacke, quartzitic sandstone and rhyolite-lavas, followed by a younger series (Dalsland) with arkose, quartzitic sandstone, slate and spilite. The formation is strongly folded along N.-S. axes, and in the more rigid basement movement along thrust planes has resulted in basement granites overriding the sediments.

The late Dalslandian granites, including the Bohus granite occurring along the coast of S.W. Sweden and continuing into S.E. Norway as the Østfold granite (Barth, 1960, 23), are younger than these sediments. This plutonism was associated with a widespread regeneration, the Sveconorwegian (see also 2.2), dated at about 1000 m.y., which affected all the older rocks of S.W. Sweden.

Eocambrian and Later

Eocambrian or Sparagmitian deposition, which certainly post-dates the Dalslandian and Jotnian, marks the beginning, in the late Precambrian, of the long Caledonian sedimentary cycle. They are a major element of the Scandinavian Caledonides (see 3.1) and of the Caledonian thrust-front. Nevertheless, the higher Sparagmitian formations, mainly quartzitic sandstones and glacogenes, spill over onto the edge of the Baltic Shield, outcropping particularly in Finnmark (Varanger Fjord) and in S.E. Norway.

On Bornholm earlier Precambrian rocks, the youngest of which are dolerite dykes, are overlain by arkosic sediments, accepted as Eocambrian.

The Visingso formation, found in the Lake Vätern basin of S. Sweden and consisting of sandstones and arkoses followed by slates and limestones, is probably equivalent to the Sparagmitian. It contains what are thought to be chitinous foraminifera, and phosphatic nodules.

Lower Palaeozoic shelf sediments, often carbonates, occur on Gotland and other islands off the S.E. Swedish coast, in isolated areas on the mainland of S. Sweden and on Bornholm. Where the Cambrian rests on the Eocambrian there is sometimes slight discordance, as on Bornholm, and sometimes no apparent break. The flat-bedding of the Lower Palaeozoic rocks emphasizes the stability of the Baltic Shield; in Scania (S. Sweden) the Lower Palaeozoic is cut by a number of N.W. faults.

Triassic, Jurassic and Cretaceous strata up to the L. Senonian occur on Bornholm, and Triassic, Rhaetic and the Danian (Palaeocene) in Scania. They show no folding, but in Scania, in particular, are cut by N.W. faults.

2.2   Southern Norway (Figs. 2.1, 2.2A, 2.3 and 2.4A)

The part of Norway described lies W. of the E. margin of the Oslo Graben and S. of the Caledonian Front. In contrast to those described above (2.1), the Precambrian rocks form high, rugged topography, intensely glaciated and rising to nearly 1500 m. There are many excellent mountain and coastal outcrops. Within the graben, N. and S. of the capital, the topography is lower, but the well-exposed Permian igneous rocks form conspicuous crags and hills.

Excursions are described by Holtedahl and Dons (1957), Barth and Bugge (1960), Dons (1960), Michot (1960) and Oftedahl (1960).

The geological succession is as follows :-
    Permian
    Lower Old Red Sandstone (Downtonian)
    Silurian
    Ordovician
    Cambrian

probably early Proterozoic
    (Telemark formation
    (Kongsberg formation
    (Bamble formation

The distinction between the different Precambrian formations is largely geographical rather than stratigraphical.

The Kongsberg and Bamble may be of the same age but their relationship is not certain. Both are separated by fault-zones from the Telemark formation of supracrustal type, but are thought to be stratigraphically older; all three seem to have shared later metamorphic events. There is also doubt regarding the age of the high-grade metasediments further S.W. in Rogaland.

The Bamble formation outcrops N.W. of the Sgagerrak and is cut off to the N.W. by a fault-zone extending S.W. for 170 km from Porsgrunn at the S.W. corner of the Oslo Graben to Kristiansand. The fault-zone is marked by mylonite and breccia.

Quartzites, greywackes, pelites, limestones and basic volcanics of the Bamble formation were laid down on a basement which can no longer be recognised. Three deformation/plutonic phases followed. The first involved folding along roughly N.-S. axes, mesazonal metamorphism and the intrusion of acid rocks (now granite-gneisses), gabbros (hyperites) and ultrabasics. The second phase occurred at deeper levels within a front of migmatisation. The structures are complex but a general N.E. trend is evident. Migmatites of great variability were formed; at a later stage, with waning orogenic stress, pegmatites (some of economic value as a source of felspar) were intruded, and also diapiric granites. This more static stage resulted in most of the earlier deformations being healed by recrystallization. Some rocks seem massive even in thin-section, the anisotropy being visible only on a larger scale (Wegmann, 1960). The last movements of the second phase were cataclastic and mark the transition into the third.

The movements of this third phase are manifest in extensive zones of mylonite, dissecting the whole region into great rhomboidal slices. These repeated movements take place at a relatively high level in a brittle environment. Some of the later deformations were probably Caledonian, Permian or even Tertiary. Although the Bamble metasediments are older than the Telemark supracrustals, the latter show the closing stages of the second phase described above and also later movements.

The Kongsberg formation lies W. of the central and northern part of the Oslo Graben and E. of a northerly mylonite/breccia zone, extending for about 100 km, which separates it from the Telemark formation. It is generally similar to the Bamble formation and has had a similar structural history. However, there is a smaller proportion of metasediments, a higher of banded gneisses, and pegmatites are rare. Rb/Sr isochrons have shown at least two metamorphic phases at about 1700 m.y. and at about 1260 m.y. Considerable mineralisation has occurred in the region, including the famous silver ores of Kongsberg mined from 1623 to 1958.

The Telemark area is noteworthy for the preservation of a large area of slightly altered metasediments which have been described as swimming in a vast sea of granites and granitic gneisses (Barth, 1960, 35). Some of the granites are regarded as formed by granitisation processes (metasedimentary structures being preserved in many cases), others to be intrusive. Long lenses of amphibolite are abundant. The best known of these is the Iveland-Ever nickeliferous amphibolite, some 50 km N. by W. of Kristiansand. This is 2-10 km wide and about 30 km long, and is extensively mined. Earlier N.W. folds and later N.S. folds have been recognised. The dominant trend in the granitic gneisses of the region is also N.-S.

The metasediments of the central Telemark area have been divided as follows:
                      Bandak Group
                      Seljord Group
                      Rjukan Group (oldest)

The Rjukan Group consists of acid lavas followed by basic lavas, with biotite schists and pebbly lavas grading into quartzites. It is separated by an unconformity from the Seljord Group made up of conglomerates, quartzites and schists. Another unconformity occurs below the Bandak Group, which contains acid and basic lavas and quartz-rich metasediments; near the top fossil algae have been found. The Seljord Group contains numerous basic sills, and small granite masses intrude the metasediments as a whole.

Although unconformities separate the three subdivisions, no evidence has been found that the oldest group has been folded three times; probably the directions of successive fold-axes are the same.

The folding, which is fairly open and has not resulted in nappe formation, etc., is, as a whole, N.-S. in the N. part of central Telemark and E.-W. further S. The northerly folding is thought to be due to the rise of the granite-gneiss to the W., and the easterly folding to the doming of the granite-gneiss to the S.

North-west and N.N.E. faults are common in both the gneisses and metasediments. A larger fault partly separates the central Telemark metasediments from the granite-gneiss to the W. This continues S.W. into Rogaland where important features are the presence of catazonal metasediments and of large anorthosite bodies (Michot, 1960).

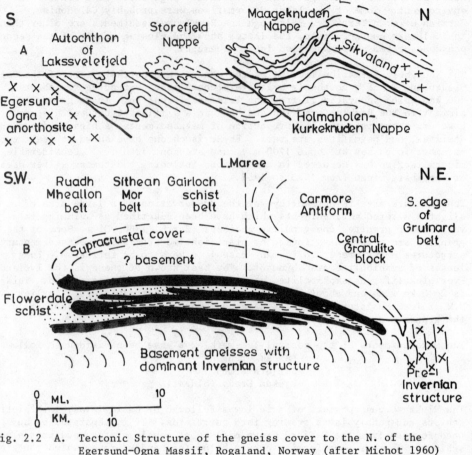

Fig. 2.2 A. Tectonic Structure of the gneiss cover to the N. of the Egersund-Ogna Massif, Rogaland, Norway (after Michot 1960)
B. Reconstruction of the Inverian structure between Ruadh Mheallon and Gruinard Bay, N.W. Scotland (after Park 1973).

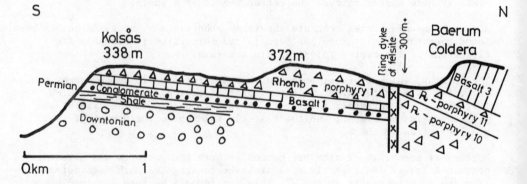

Fig. 2.3 South margin of Baerum Caldera, Oslo Graben, Norway (after Holtedahl and Dons 1957)

The structures of the Archean rocks of Rogaland are well-exposed N.N.E. of Egersund, where there is an autochthon of gneiss forming part of the framework of the Egersund-Ogna anorthosite.

Most of this part of the autochthon consists of grey, strongly-banded gneisses with hypersthene (charnockite), interpreted as formed by the granitic migmatisation of norite. The gneissose structure dips N.E. at about $45°$ and the structures include a syncline overturned towards the S.W. and pitching S.E. In monoclinal folds, which have gaped, there are phacoliths of massive granite-gneiss with orthoclase augen. This autochthon is overridden by a nappe which is well seen in the mountain of Storefjeld, N. of Vigesa, as a recumbent fold. The lower inverted limb is seen near the foot of the mountain in migmatite with dark noritic bands with boudinage structure, and the upper limb at the top of the mountain. The fold is also brought out by a massive granite gneiss above and below the noritic migmatite.

North-west of Vigesa, near the E. end of the lake of Holmvand, paragneisses are seen including quartzite, diopside-gneiss and cordierite-sillimanite-gneiss. These make it possible to work out the structure in detail. There was first the development of N.-S. folds, which was followed by the intrusion of the Egersund-Ogna anorthosite, to be described in a moment. Secondly, there was the main recumbent folding with nappe formations along E. to S.E. axes, and thirdly, local, open E.-W. folds.

The Egersund-Ogna anorthosite, already referred to, forms a great dome in the autochthon just described. Between the anorthosite and the cover on the N. margin norite has been intruded. Near the margins, owing to later dynamic metamorphism, the anorthosite is gneissose and the norite streaked out.

East of the Egersund-Ogna mass, occupying a syncline curving parallel to the margin of the dome, lies the Bjerkreim-Sogndal complex. In general this has a framework of norite and leuconorite showing strongly marked, graded bands, up to about 0.70 m, with a leuconorite top and a base rich in dark minerals which have every appearance of being due to settling out from convection currents. In the centre, forming a high bare plateau, is quartz-monzonite.

At the S.E. end of this complex, near Sogndal, is a worked ilmenite-magnetite ore body with about 18% TiO_2. The readily workable portion of the body is approaching exhaustion and future development will be in a body recently discovered by aerial magnetic survey at Tellenes.

The metamorphism in Rogaland is interpreted as monocyclic and dated (Michot and Pasteels, 1968) at some 1000 m.y. The emplacement of the mangerite of the Bjerkreim-Sogndal mass is dated at 950 m.y. Both the catazonal metamorphism in Rogaland and the mesozonal metamorphism in Agder to the S.E. thus belong to the Sveconorwegian cycle.

Lower Palaeozoic sediments overlie the Precambrian N. and S. of Stavanger.

In the Fen area, 120 km S.W. of Oslo, the gneisses are penetrated by peralkaline and carbonatite rocks of peculiar composition. It was thought these might be related to the Permian igneous rocks of the Oslo region, but age-determinations, although variable, give Eocambrian to Lower Palaeozoic dates (Bèrgstol and Svinndal, 1960).

The Oslo Graben is about 200 km long in a N. by E. direction and from 30 to
70 km wide. On its E. margin are the Precambrian rocks of S.E. Norway (2.1),
and on its W. margin are those of S. Norway just described. Precambrian
rocks come up within the Graben in the horst forming the Nesodden peninsula,
S. of Oslo, and on the W. side of the Oslo Fjord, S.W. of the city. In the
Nesodden peninsula there is a dome of granite-gneisses, garnetiferous
amphibolites and pegmatites mantled through a transition zone of migmatites
by psammitic and pelitic metasediments. Kyanite and staurolite occur in the
pelites; extreme distortion and boudin structures suggest great plasticity.

On the W. side of the fjord there is a basal conglomerate of Middle Cambrian
age resting on the peneplaned Precambrian. This is followed by 50-60 m of
alum-shale with lenses of dark limestone containing fossils. Lower, Middle
and Upper Ordovician are all represented: the Lower by graptolitic shale
and limestone beds with shelly fauna, and the Middle and Upper mostly by
nodular limestones and dark calcareous shales with limestone lenses; some
sandy beds also occur. The Silurian mainly consists of calcareous and
arenaceous shales and of limestones.

West of Oslo, Upper Llandovery limestones with corals are overlain by
Downtonian non-marine sandstone (Ringerike Sandstone), about 500 m thick,
consisting of hard reddish and grey sandstone with ripple marks. Further N.
sedimentation started in Lower Cambrian or even Eocambrian times.

Although these Lower Palaeozoic rocks apparently lie S. of the Caledonian
Front they differ from those of the Baltic Shield (2.1) in being markedly
affected by Caledonian folding which must be of late date as the Downtonian
is also involved. The folding is intense in the N. of the Graben, moderate
around Oslo (where N.E. flexures, overturned to the S.E., are well-displayed)
and dies out S. of Drammen. It thus clearly wanes away from the Caledonian
Front. It is possible that reverse faults, common in the Oslo region and
further N.W., represent imbricate structures associated with a decollement
thrust plane underlying the northern part of the Oslo Graben. If this is the
case, the Caledonian Front may well lie farther S. than has been supposed.

Graben development in early Permian times was accompanied by the uprise of
magma, mainly subalkaline and intermediate to acid in composition. The lavas
are rhomb-porphyries with a few basalts and in places rest on thin fossiliferous Permian sediments of Rotliegendes type. At least six of the intrusive
centres form ring-complexes with diameters ranging from 6 to 16 km; a
tectonic aspect, namely subsidence, is therefore involved. The most striking
is the Baerum Caldera, N.W. of Oslo.

Volcanic activity started with a basalt flow, then came nine rhomb-
porphyries, then another basalt, then two or three rhomb-porphyries, then
several basalts and finally more rhomb-porphyries.

On Kolsås, a conspicuous hill N.W. of Oslo, Downtonian, almost flat-lying in
the centre of a syncline, is overlain by gently-dipping red Permian fossiliferous siltstone. Above come about 3 m of quartz-conglomerate immediately
overlain by basalt. Above this comes the lower rhomb-porphyry, forming the
summit of Kolsås. To the N. the porphyry comes against a thin dyke of
felsite. The felsite is rich in pyrites and brecciated; brecciation is also
seen in the flanking rocks. Another porphyry occurs to the N., but this is
number eleven in the sequence and is followed a short distance to the N. by
the overlying basalt. Thus there is a difference in stratigraphical level of

Fig. 2.4A. Boudinaged pegmatite in Archean metasediment, Oslo Fjord, Norway.

B. Unconformity of Torridonian gritty arkose on Lewisian North Shore of Loch Assynt, Sutherland, Scotland.

over 300 m between the lavas N. and S. of the felsite which must, therefore, follow a fault. The fracture, in fact, is followed by a ring-dyke enclosing an area of the higher lavas about five miles in diameter.

Permian basic dykes are abundant in the Oslo region and some of the basic dykes which occur in the Precambrian as far W. as Rogaland may be of the same age.

Further graben faulting followed the Permian igneous activity and movement may have gone on until the Tertiary.

2.3 North-West Scotland (Figs. 2.2B, 2.4B and 2.5)

The Precambrian Block, which lies W. of the Caledonian Front in Scotland (the Moine Thrust Zone, 3.2), is small compared with the Baltic Shield. It is, however, a remnant of a much larger block (sometimes called Eria), part of which is drowned and part of which lies on the other side of the Atlantic Rift; a remnant occurs around Rockall (8.11 and 9.2).

The Precambrian Blocks form a long outcrop on the mainland of the N.W. Highlands extending from Cape Wrath to Loch Alsh, a sparsely inhabited region deeply indented by fjord-like sea lochs. The Lewisian "basement" largely makes bare, hilly ground with small lochs and peaty hollows. Above, the spectacular relic mountains of Late Precambrian (Torridonian) sandstone rise to 3483 ft (1062m), and are sometimes capped by Cambrian quartzite. The Precambrian also appears in a number of islands of the Inner Hebrides and forms the long island chain of the Outer Hebrides, including Lewis, the type locality. Here, the gneisses reach a height of 2622 ft (799 m).

A number of excursions are described by Macgregor and Phemister (1972).

The succession is as follows :-
                      Tertiary
                      Jurassic
                      Triassic
                      Ordovician
                      Cambrian

Upper Proterozoic	(Upper Torridonian ((((Lower Torridonian	(Aultbea Group (Applecross Group (Diabaig Group Stoer Group	760 m.y. 995 m.y.
Lower Proterozoic		(Laxfordian cycle ((Inverian cycle (	1500-1600 m.y. 2200 m.y.
Archean	(Lewisian	(Scourian cycle	2400-2800 m.y.

No regional metamorphism and only gentle folding took place after the Laxfordian cycle. This contrasts with the Sveconorwegian metamorphism (1000 m.y.) and the strong foreland folding of parts of the Baltic Shield (2.1 and 2.2).

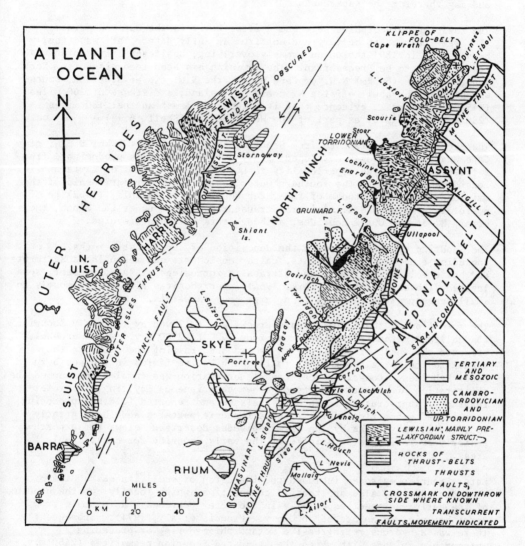

Fig. 2.5 STRUCTURE OF THE NORTH-WEST KRATOGEN OF SCOTLAND

Archean and Lower Proterozoic Evolution

No isotopic dates significantly older than 2800 m.y. have so far been recorded; nevertheless, as discussed by Watson (1975, 19), some gneisses may be older than supracrustal rocks affected by the 2800 m.y. metamorphism and may therefore be Katarchean.

The Lewisian metamorphics were named after Lewis in the Outer Hebrides; they were thought to be very ancient from an early date. This was finally proved when in 1891 the unconformably overlying, unaltered Torridonian was itself shown to be Precambrian. The Lewisian has been recognised from North Rona, 40 miles (64 km) N.W. of Cape Wrath, the N.W. tip of Scotland, southwards to Inishtrahull off the N. coast of Ireland, a distance of 260 miles (415 km). There is evidence that it extends at least another 180 miles (290 km) to the N.E. as part of the West Shetland Shelf (8.11).

The Scourian takes its name from Scourie, about 20 miles (32 km) S.S.W. of Cape Wrath. Structures developed mainly during this cycle predominate from a few miles N. of the type-locality to Loch Maree, in parts of Lewis and in considerable areas in the southern Outer Hebrides. The relationship of the mainland outcrops to those of the Outer Hebrides led Dearnley (1962) to postulate a major N.N.E. sinistral transcurrent fault under the Minch, the strait E. of the Outer Hebrides, although this has been disputed.

The Scourian cycle opened with the deposition of supracrustal rocks, including fine clastic sediments, calcareous cherts and ferruginous sediments and basic volcanics. Basic and ultrabasic intrusions are also present, some layered, (Bowes, Wright and Park, 1964). Anorthosites also occur, mostly in small lenses, although there is a large body in Harris.

The formation of these supracrustal rocks was followed by the early Scourian (Badcallian) metamorphism, dated at from 2900 to 2700 m.y., which probably affected the whole of the Lewisian as exposed, raising the rocks to the granulite facies, although only to the amphibolite facies in the southern Outer Hebrides. Coeval or near-coeval deformation was complex and caused the obliteration of primary structures other than igneous layering, the interleaving of supracrustal rocks with gneisses and repeated folding, producing complicated geometrical forms. These tectonic patterns have been greatly modified in many areas by the later episodes described below, but a northeasterly grain is characteristic of the early Scourian for considerable areas.

Later Scourian episodes included the intrusion of small igneous bodies of basic to acid and also anorthosite composition, which locally cut the banding of the early gneisses, especially in the Outer Hebrides; an example from Barra has yielded an age of 2600 m.y. (Francis et al., 1971). Subsequently, there was a hydrous retrogressive metamorphism during or preceding a metamorphic episode with which the age-dated Scourian pegmatites (2460 m.y.- Giletti et al., 1961) are genetically related. Geniculate folding also affected the rocks about this time. Migmatisation and amphibolite facies metamorphism also took place accompanied by ductile deformation.

The Loch Maree Group of metasediments and metavolcanics includes mica-schists, quartz-schists, calc-silicate rocks, limestones and kyanite-schists; magnetite-banded ironstones also occur, as in so much of the Precambrian, but not in workable quantities. There are also thick sheets of metabasics which are regarded as intrusive into the metasediments. These show no sign of the early granulite facies metamorphism, and Bowes (1968) and Park (1970) have suggested that they accumulated between the Badcallian phase of the Scourian cycle and the Inverian cycle.

The Inverian (named after Loch Inver) was marked by a series of metamorphic events, often of amphibolite-facies, and deformational events including north-westerly monoclinal folding. It may have begun with the intrusion of a pegmatite suite as early as 2500-2400 m.y. and closed with the intrusion of a dyke swarm under static conditions about 2190 m.y. ago. These north-westerly dykes are dominantly basic and ultrabasic and cut the Scourian and earlier Inverian structures.

In zones which have escaped Laxfordian metamorphism the dykes are very little altered and even ophitic structure is preserved; where the dykes pass into the zones of Laxfordian metamorphism they become progressively amphibolitized and schistose. Some authors hold that the dykes were intruded as a large swarm over a single time-span. Bowes and Khoury (1965) have shown that early dykes have suffered deformation and metamorphism prior to the emplacement of later dykes, and Hopgood (1971) has identified dykes deformed by folds equated with structures which are cut by dykes at another locality. Park and Cresswell (1972) have stated that there is no evidence of two or more swarms separated by tectonic episodes.

The Laxfordian cycle, named after Loch Laxford in the N.W. of the mainland, produced the greater part of the Lewisian structure as now seen. Apart from the larger zones, from which the Scourian and Inverian events described above have been determined, many small masses with distorted earlier structures may be distinguished within the Laxfordian complex.

Laxfordian deformation and metamorphism is clearly polyphase, but variations in radiometric data suggest that successive phases in different districts may have been diachronous.

For parts of the mainland an early phase of granulite-facies metamorphism, accompanied by repeated deformation including thrusting and possible nappe formation, has been proposed. Granulite-facies metamorphism has also been suggested by Dearnley (1962) for an early phase in the Outer Hebrides. Others (Coward et al., 1969) have put forward the view that early Laxfordian metamorphism was mainly of amphibolite-facies.

Main phase movements were characterised by N.W.-S.E. trending major folds, with steep axial planes.

Subsequent to the main phase movements north-westerly dolerite dykes were emplaced. These were followed by pegmatite injection and by late phase movements accompanied by low-grade metamorphism. In a broad sense the earlier episodes seem to have been marked by "dry" metamorphism and the late phase by "wet" metamorphism in amphibolite-facies and by extensive migmatisation and pegmatite injection.

In the Outer Hebrides Hopgood and Bowes (1972) have distinguished no less than six Laxfordian episodes based on minor structures.

The youngest date recorded for a Laxfordian event is 1150 m.y.

At many localities in the Lewisian thin belts of flinty-crush rock and mylonite occur, frequently with N.W. trend. Some may be contemporaneous with the north-westerly Laxfordian folding, others may be later, although they are certainly pre-Torridonian.

A wide belt of mylonite and flinty-crush rock, inclined at low angles to the E.S.E., skirts the eastern coasts of the Outer Hebrides. Sheared and mylonitized gneisses are intruded by flinty-crush rock; the latter generally occurs in thin veins but bands up to 100 ft (30 m) thick have been recorded. The zone is held to mark the course of a pre-Torridonian thrust-zone.

Upper Proterozoic and Later Evolution

Even if the latest Laxfordian date is accepted, some 150 m.y. elapsed before the deposition of the Lower Torridonian, on a land surface with relief of up to 1312 ft (400 m).

The Stoer Group, confined to a comparatively small outcrop in the Stoer Peninsula, consists of hard, fairly steeply-dipping, red sandstones, with a mudflow horizon containing volcanic material. Its separation by an unconformity from the Upper Torridonian was not recognised until about 1967. That the unconformity represents a big time-interval has been shown by a radiometric age for the Lower Torridonian of about 995 m.y. and for the Upper Torridonian Applecross Group of about 761 m.y. This is in accord with the very different palaeomagnetic pole positions reported for the Lower and Upper Torridonian.

The Stoer Group, which shows dips of $25°$ to $30°$ to the W., was subjected to marked deformation before the deposition of the Upper Torridonian. This unconformably overlies the Stoer Group but rests mainly on the Lewisian with a strongly diachronous base, frequently marked by breccia or conglomerate of local origin. At some localities, notably Slioch on the E. side of the Loch Maree, the Torridonian abuts against hills of Lewisian gneiss, which still rise up to 2000 ft (610 m) above the general gneiss surface.

The main subdivisions of the Upper Torridonian are :-
                    Aultbea Group
                    Applecross Group
                    Diabaig Group.

The Diabaig Group consists of red sandstones, red mudstones and grey, sandy shales. It thins out to the N., where the Applecross, deposited in several alluvial fans, rests on an ancient pediment of Lewisian gneiss (Williams, 1969). The Group in general consists of red arkoses and pebbly conglomerates, with current-bedding indicating origin from the W. The pebbles include exotic supracrustal metasediments and acid extrusives which can be matched in Greenland. The Aultbea Group consists of sandstones, flags and grey shales; Nannofossils (Downie, 1962) from the Upper Torridonian support a late Precambrian age.

Red sandstones and conglomerates resting on the Lewisian N. of Stornoway in Lewis have been mapped as Torridonian, but are regarded by others (e.g. Steel, 1971) as Triassic.

Successions correlated with varying certainty with subdivisions of the Lower or Upper Torridonian occur in Hebridean islands from Skye to Islay. As nearly all of these, however, occur within the Caledonian frontal thrust-zone, they are dealt with in the next chapter (3.1).

West of the thrust-zone the Torridonian strata have low dips. Nevertheless, the overlap of Cambrian on to Lewisian shows that Precambrian folding took place. In particular, an upfold developed N. of Assynt, where Cambrian rests on Lewisian. In Assynt the Torridonian was inclined to the W. at $10°$ to $20°$, to be restored to the near horizontal by post-Lower Ordovician tilt to the E. of the same amount. In the occurrence of these late Precambrian movements the Scottish Shield contrasts with the Baltic Shield (2.1) where the Eocambrian generally passes up conformably into the Cambrian.

In the North-West Highlands the Cambrian Basal Quartzite rests unconformably on an almost planar Lewisian/Torridonian surface. It is followed by the quartzitic Pipe-rock, named after the sand-filled tubes it contains, interpreted as worm-burrows. The Fucoid Beds, erroneously named after supposed seaweed markings (probably worm-casts), consist of dolomitic shales and siliceous dolomites. They contain the trilobite Olenellus. The Serpulite Grit is a gritty quartzite with Salterella. The Durness Limestone, up to 2000 ft (612 m) thick, is largely dolomitic. It has Lower Cambrian fossils not far above the base and an abundant gastropod/cephalopod/trilobite fauna of Lower Ordovician age in its uppermost subdivisions.

Gentle folding, including the restoration of the Torridonian of Assynt to the near horizontal mentioned above, affected the Scottish foreland. The folding may well have been late-Caledonian as it deforms the border-thrusts (3.1) now known to be Lower Devonian.

Numerous faults cut the region, dominant directions being N.E. to N.N.E. and N.W. to W.N.W. In part, at least, they are wrench-faults. Several of the north-westerly faults cut across the thrust-zone and continue into the orogen. The greatest of the north-westerly fractures is the Loch Maree Fault along which dextral transcurrent movement of about four miles has taken place. Vertical, as well as lateral, displacement along such faults has occurred and the combination can be deduced in the thrust-belt where strata and thrust-planes with different, and often opposing, dips are common.

The Loch Maree Fault, and probably others, lies along a zone of weakness initiated in pre-Torridonian times. Some of the movements along the faults of the region may, like the folds, be late-Caledonian, some Hercynian. Moreover, the displacement of Triassic red beds, Liassic limestones and Eocene basalt-lavas show that faulting continued into the Tertiary.

REFERENCES

Barth, T.F.W. 1960a. In: The Geology of Norway. (ed. O. Holtedahl). Oslo.
Barth, T.F.W. 1960b. Precambrian Gneisses and Granites of the Skagerak Coastal Area, South Norway. Internat. Geol. Congr. (Norden), Exc. Guide A8.
Barth, T.F.W. & Bugge, J.A.W. 1960. Precambrian Gneisses and Granites in the Skagerak Coastal Area, South Norway. Internat. Geol. Congr. (Norden), 1960, Exc. Guide A5.
Bergstol, S. & Svinndal, S. 1960. In: The Geology of Norway. (ed. O. Holtedahl). Oslo.

Bowes, D.R. 1968. An orogenic interpretation of the Lewisian of Scotland. Intern. Geol. Congr., 4, 225-236.
Bowes, D.R. & Khoury, S.G. 1965. Successive periods of basic dyke emplacement in the Lewisian Complex south of Scourie, Sutherland. Scot. Journ. Geol., 1, 295-299.
Bowes, D.R., Wright, A.E. & Park, R.G. 1964. Layered intrusive rocks in the Lewisian of the north-west Highlands of Scotland. Q.J.G.S., 120, 153-184.
Bugge, J.A.W. 1960. Precambrian of eastern Finnmark. In: The Geology of Norway. (ed. O. Holtedahl). Oslo.
Coward, M.P. et al. 1969. Remnants of an early metasedimentary assemblage in the Lewisian Complex of the Outer Hebrides. Proc. Geol. Assoc., 80, 387-408.
Dearnley, R. 1962. An outline of the Lewisian Complex of the Outer Hebrides in relation to that of the Scottish Mainland. Q.J.G.S., 118, 143-176.
Dons, J.A. 1960a. In: Geology of Norway. (ed. O. Holtedahl). Norges Geologiske Undersøkelse, 208.
Dons, J.A. 1960b. Stratigraphy of supracrustal rocks, granitization and tectonics in the Precambrian Telemark area. Internat. Geol. Congr. Exc. Guide A10.
Downie, C. 1962. So-called spores from the Torridonian. Proc. Geol. Soc. London, 1600, 127-128.
Francis, P.W. et al. 1971. Isotopic age dates from the Isle of Barra, Outer Hebrides. Geol. Mag., 108, 13-22.
Geijer, P. 1963. The Precambrian of Sweden. The Precambrian, vol. 1. (ed. K. Rankama). New York.
Giletti, B. et al. 1961. A geochronological study of the metamorphic complexes of the Scottish Highlands. Q.J.G.S., 117, 233-264.
Holtedahl, O. & Dons, J.A. (eds.). 1957. Geological Guide to Oslo and District. Oslo.
Hopgood, A.M. 1971. Structure and tectonic history of the Lewisian gneiss, Isle of Barra, Scotland. Krystalinikum, 7, 27-60.
Hopgood, A.M. & Bowes, D.R. 1972. Application of structural sequence to the correlation of Precambrian gneisses, Outer Hebrides, Scotland. Bull. Geol. Soc. Am., 83, 107-128.
Internat. Geological Congress (Norden), 1960. Excursion Guides A21, A22, A23 and A24.
Macgregor, M. & Phemister, J. 1972. Geological Excursion Guide to the Assynt District of Sutherland (3rd ed.). Edin. Geol. Soc.
Magnusson, N.H. 1940. Ljusnarsbergs malmtrakt (Geology and ore deposits of Ljusnarberg). Sveriges. geol. Unders., Ser. Ca30.
Magnusson, N.H. 1965. The Precambrian history of Sweden. 19th William Smith Lecture. Q.J.G.S., 121, 1-30.
Michot, P. 1960. La géologie de la catazone: Le problème des anorthosites, la palingénèse basique et la tectonique catazonale dans la Rogaland meridional Norvège meridionale. Internat. Geol. Congr. (Norden), Exc. Guide A9.
Michot, J. & Pasteels, P. 1968. Etude Géochronologique du Domaine Métamorphique du Sud-Ouest de la Norvège. Ann. Soc. Géol. Belgique, 91, 93-110.
Oftedahl, I.W. et al. 1960. The Larvik - Langesund and the Fen Areas. Internat. Geol. Congr. (Norden), Exc. Guide A12.
Padget, P. 1973. Evolutionary Aspects of Precambrian of Northern Sweden. In: Arctic Geology. (ed. M. G. Pitcher). Tulsa.
Park, C.F. & MacDiarmid, R.A. 1970. Ore Deposits. (2nd ed.). San Francisco.
Park, R.G. 1970. Observations on Lewisian chronology. Scot. Journ. Geol., 6, 379-399.

Park, R.G. 1973. The Laxfordian belts of the Scottish Mainland. The Early Precambrian of Scotland and related rocks of Greenland, 68. Univ. of Keele.

Park, R.G. & Cresswell, D. 1972. Basic dykes in the early Precambrian (Lewisian) of N.W. Scotland: their structural relations, conditions of emplacement and orogenic significance. Intern. Geol. Congr., 1, 238-245.

Steel, R.J. 1971. New Red Sandstone movement on the Minch fault. Nat. Phy. Sci., 234, 158-159.

Watson, J. 1975. The Lewisian Complex. Geol. Soc., Spec. Rep. No. 6, 15-29.

Wegmann, E. 1960. In: Geology of Norway. (ed. O. Holtedahl), 6. Norges Geologiske Undersøkelse, 208.

Welin, E. & Lundqvist, T. 1970. New Rb-Sr age data for the sub-Jotnian volcanics (Dala porphyries) in the Los-Hamra region, central Sweden. Geol. Foren. Stockholm Forh., 92, 35-39.

Williams, G.E. 1969. Characteristics of a Precambrian pediment. Journ. Geol., 77, 183-207.

Chapter 3
CALEDONIAN FOLD-BELTS
(MAINLY METAMORPHIC)

3.1 Introduction

A well-defined Thrust-Belt separates the Precambrian Block (2.3) of North-West Scotland (the Hebridean Kratogen) from the sector of the Caledonides which underlies most of the British Isles and extends southwards to the Midlands Kratogen (6.4) and the Hercynian Front (5.11 and 6.5). Within the British Caledonides all pre-Devonian rocks N.W. of the Highland Boundary Fracture-Zone (3.3 and 3.4) have undergone metamorphism of regional type. South-east of this Fracture-Zone (apart from Western Ireland (3.5)) Caledonian metamorphism has not gone beyond the production of slatey cleavage. In Scotland the Caledonides of the Northern Highlands (3.3) are separated from those of the Grampian Highlands (3.4) by the Great Glen Fault (3.3).

The Orkney Islands are a continuation of the Northern Highlands, but the Shetland Islands (3.6), while they can be related to the Highlands, also form a stepping-stone to Scandinavia.

In Scandinavia a Thrust-Zone (3.7) separates the Baltic Shield (2.1) and the Precambrian Block of Southern Norway (2.2) from the Caledonides (3.8). This consists mainly of metamorphics, but also contains considerable areas of unaltered Upper Proterozoic and Lower Palaeozoic sediments, which, however, are not as sharply separated from the metamorphics as in the British Isles. Post-tectonic plutonics are widely distributed throughout both the metamorphic and non-metamorphic Caledonides.

3.2 The North-West Caledonian Front (Figs. 2.5 and 3.1)

Thrust structures separating the N.W. kratogen from the highly folded and metamorphosed rocks of the Caledonian Orogen are spectacularly displayed on the bare rock faces of deeply dissected mountains.

The superposition of altered on unaltered rocks in North-West Scotland caused early controversy, which became acute when Charles Peach's (1854) discovery of Cambrian fossils in the foreland strata made it clear that Lower Palaeozoic sediments are overlain by highly metamorphosed rocks. The problem was solved by Lapworth's (1883) demonstration that thrusting, accompanied by the development of mylonite, had taken place on a large scale. This view was confirmed by detailed mapping of the region (Peach and Horne, 1884; Peach et al., 1907), which includes the famous Assynt district.

Excursion itineraries have been given by Macgregor and Phemister (1972) and by Barber and Soper (1973).

In stratigraphical order the formations involved in the Thrust-Zone are as follows:-

CALEDONIAN - METAMORPHIC

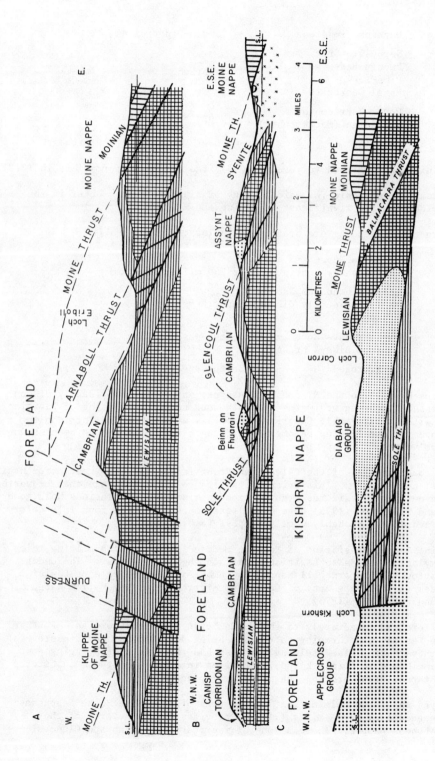

Fig. 3.1 SECTIONS OF NORTH-WEST CALEDONIAN FRONT
A, Loch Eriboll - Durness; B, Assynt; C, Loch Carron - Loch Kishorn

Durness Limestone	– Cambro-Ordovician
Serpulite Grit)	
Fucoid Beds)	– Cambrian
Pipe-rock)	
Basal Quartzite)	
Upper Torridonian = Moinian (metamorphic)	– Upper Proterozoic
Lewisian	– Lower Proterozoic and Archean

The Moinian is described in (3.3 and 3.4), the remaining formations in (2.3).

Hypabyssal and plutonic intrusions, many of alkaline types, occur in the Thrust-Zone and are significant in dating the movements.

The structural elements are broadly as follows:-

    Moine Nappe – exotic
    (probably continuous for 250 miles; 400 km)

    Autochthonous Nappes
    (individually discontinuous; various local names)

    Imbricate Zone – autochthonous
    (not always present)

    Kratogen – autochthonous

From Loch Eriboll on the N. coast to Assynt, major but individually discontinuous thrusts bring westwards autochthonous nappes consisting of Lewisian slices with unconformable Cambrian. From Assynt southwards, Torridonian also appears, and folding becomes significant; dislocation has been shown to be polyphase.

The N. coast section of the Caledonian Front is not only of great structural interest but is also of historical significance as it was here that Lapworth carried out his classic research. Although his broad conclusions hold good, Soper and Wilkinson (1975) have shown that there have been four fold deformation sequences, probably wholly of post-Cambrian age.

The Moine Nappe, consisting of Moinian metasediments (3.3) overlying thick blastomylonites and also Lewisian, rests on the Moine Thrust. The underlying autochthonous Arnaboll Nappe, of Cambrian unconformable on Lewisian, overrides an imbricate zone of Cambrian referred to as the Heilem Nappe by Soper and Wilkinson (1975).

In the foreland immediately to the W. Cambrian overlies Lewisian. Further W. step-faults, downthrowing to the W., bring down the Durness Limestone of the Durness Basin. The faulting has also had the effect of bringing down the Moine Nappe to form a klippe, ten miles in advance of the main outcrop of the thrust, thus proving the minimum displacement.

In Assynt an eroded culmination, on the axis of which the Moine Nappe has been denuded back seven miles (11 km), provides a wide outcrop of the autochthonous nappes. Above a zone of imbricated Cambrian the Glencoul

Thrust brings forward a 1600 ft (490 m) thick slice of Lewisian overlain by Cambrian. In the overlying Assynt Nappe the Lewisian has a cover of both Torridonian and Cambrian which are recumbently folded; this nappe is split by the Ben More Thrust.

At the S. end of the culmination the Assynt Nappe is overridden by the Moine Nappe bringing Moinian metasediments on to the Cambrian Limestone of the foreland. The thrusts have themselves been folded, and klippen occur in eroded synclines.

Spectacular exposures of the front occur between the S. end of Loch Maree and Loch Alsh. Cambrian Basal Quartzite and Pipe-rock, forming 3309 ft (1010 m) Beinn Eighe, are overlain to the E. by Torridonian brought forward by the Kinlochewe Thrust.

The Kinlochewe Nappe in its turn is overridden to the E. by the Moine Thrust. North-east of Beinn Eighe, however, the Cambrian rocks of the foreland are overlain by a mass of Torridonian forming an outlier or klippe of the Kinlochewe Nappe $3\frac{1}{2}$ miles in advance of the main outcrop.

Further S. the structural continuation of the Kinlochewe Nappe is known as the Kishorn Nappe. The main part of this nappe consists of a great recumbent syncline, closing towards the E. and pitching N. above an imbricate zone. Near Loch Kishorn imbricated Cambrian is faulted down to the W. against the Torridonian of the Applecross Peninsula. Above the imbricate zone, E. of the Loch, the Kishorn Thrust brings forward the Diabaig Group of the Torridonian overlain still further E. by the Lewisian. This inverted limb of the recumbent syncline continues as far S. as Loch Alsh. Along the N. shores of the latter, E. of Kyle, the Applecross Group, with inverted current-bedding, passes under the Diabaig Group. This in its turn is overlain along the Balmacara Thrust by Lewisian, above which comes Moinian and Lewisian of the Moine Nappe. The overfolding is earlier than the thrusting; in fact, eastwards from Kyle axial-cleavage is replaced by lower-angled thrust-cleavage. In Skye, the lower limb of the syncline is seen, consisting of current-bedded Torridonian in normal order. Here the Kishorn Thrust brings the Applecross Group over Durness Limestone, thermally metamorphosed and structurally distorted by Tertiary intrusion. Further S. anticlinal folding and erosion of the Kishorn Nappe has revealed the underlying foreland in the Window of Ord.

At the S. end of the Sleat Peninsula the three Tarskavaig Nappes intervene between the Moine and Kishorn Nappes. The Tarskavaig Nappes consist of gritty felspathic sandstones, strongly granulitized, which contain metamorphic biotite and small manganese garnets.

South of Skye the Caledonian Front is largely hidden under the sea and obscured by Tertiary intrusions. The Loch Skerolls Thrust in Islay is probably the continuation of the Moine Thrust displaced by the Great Glen Fault (3.3). Most or all of the Lewisian and Torridonian rocks of the Inner Hebrides, to judge from their structure, occur within the Thrust-zone.

The island of Innishtrahull, off the N. coast of Ireland, consists of Lewisian rocks, and it seems probable, therefore, that the continuation of the Moine Thrust passes between the island and the mainland of Ireland. If this is accepted, the Caledonian Front becomes traceable for 250 miles in the

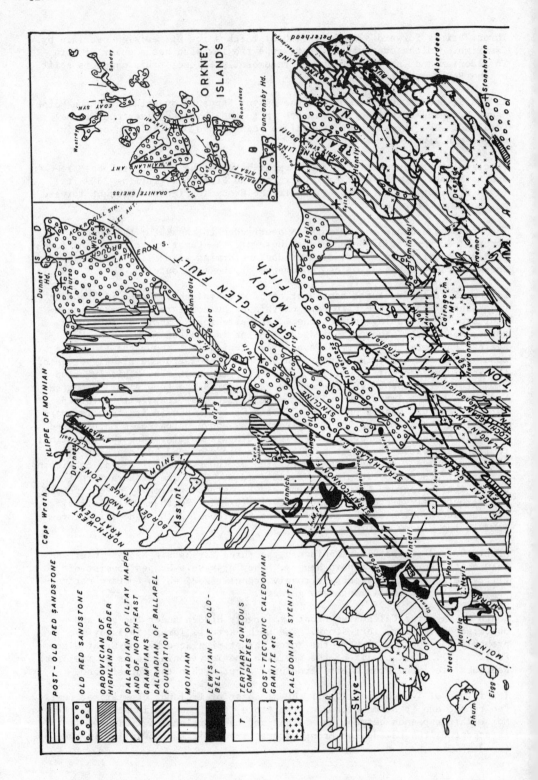

CALEDONIAN - METAMORPHIC

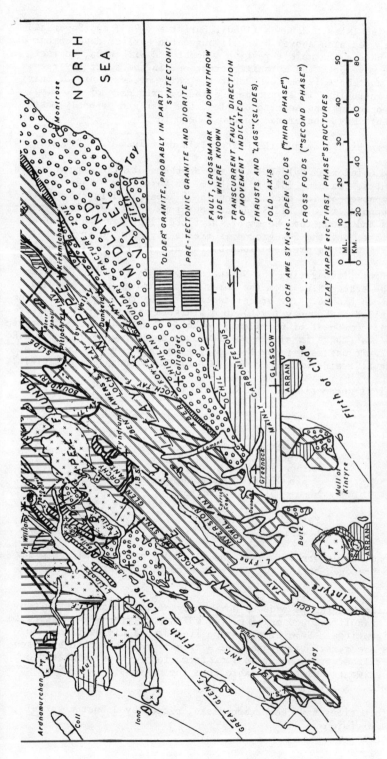

Fig. 3.2 STRUCTURE OF THE NORTHERN AND GRAMPIAN HIGHLANDS OF SCOTLAND
F.W.S. = Fort William Slide; I.B.S. = Iltay Boundary Slide;
I.F. = Inninmore Fault; L.M.F. = Loch Maree Fault;
L.S.T. = Loch Skerrols Thrust

N.W. of the British Isles. Johnson (1960), Christie (1963) and Soper and Wilkinson (1975; see also above) have all suggested earlier development of mylonitic rocks before the main thrusting. Soper and Wilkinson, in fact, consider all the movements and mylonite formation, in the N. at any rate, to be post-Arenig, and there is no doubt that the main movements are of this date. Alkaline intrusions in Assynt have a radiometric date of 400 m.y. As the thrusting and intrusions seem to overlap, the last and probably most important movements may, therefore, be Lower Devonian.

3.3   The Northern Highlands and the Orkney Islands (Figs. 3.2, 3.3 and 3.4)

The rocks of the Caledonian Fold-Belt in the Northern Highlands form rugged, mountainous country, reaching 3873 ft (1180 m) at Carn Eige. There are, however, wide expanses of low ground along the E. coast and in the Orkney Islands, corresponding to Old Red Sandstone outcrops. Good exposures occur in the mountains and along the shores of the fjord-like, sea-lochs which incise the W. coast, but the low ground is frequently mantled in glacial deposits. Some excursions are described by Lambert and Poole (1964).

The Great Glen Fault, separating the Northern Highlands from the Grampian Highlands, can be traced S.W. from the Moray Firth through Loch Ness and other lochs to Loch Linnhe. The brecciated zone is up to a mile wide; the fault was recognised by Kennedy (1946) as a sinistral transcurrent fracture with a horizontal component of 65 miles (104 km). Comparison of metamorphic levels suggests that there is also a considerable downthrow towards the S.E. The Great Glen Fault passes off the N.W. of Ireland and may continue as a lineament on the S.E. side of the Porcupine Bank (Riddihough and Max, 1976).

The stratigraphical succession of the Northern Highlands is:-

>   Tertiary
>   Cretaceous
>   Jurassic
>   Triassic
>   Carboniferous
>   Upper Old Red Sandstone
>   Middle Old Red Sandstone
>   Lower Old Red Sandstone
>   Moinian
>   Lewisian

The Lewisian consists of orthogneisses and metasediments generally similar to those of the kratogen to the W. (2.3) but affected by Caledonian events (see below).

Lithologically the Moinian is made up dominantly of meta-psammites, often referred to as granulites, and meta-pelites which vary from mica-schists to gneisses and migmatites. Some of the granulites are highly siliceous but true quartzites are rare. Limestones are extremely rare but thin beds and lenses of calc-silicate rock occur in some formations. Some hornblendic rocks are metasediments, distinguishable from the metabasic intrusions which are also present.

The following formations have been recognised, placed by Johnstone et al., (1969) in three divisions:-

Loch Eil Psammite	Loch Eil Division
Glenfinnan Striped Schist) Lochailort Pelite)	Glenfinnan Division
Upper Morar Psammite) Morar Striped and) Pelitic Schist) Lower Morar Psammite) Basal Pelite)	Morar Division

The junctions between the divisions are, in places at any rate, tectonic (see below); nevertheless, it is possible that the above table represents a stratigraphical succession.

The Moinian was intruded prior to metamorphism and most of the folding (see below) by the Carn Chuinneag-Inchbae Granite, and a considerable number of post-tectonic Caledonian plutons occur.

The Old Red Sandstone (Devonian) rests with strong unconformity on the Moinian metamorphics and the granites. In Caithness, and round the Dornoch and Cromarty Firths, Lower, Middle and Upper Old Red Sandstone have been recognised. In Orkney the Middle Old Red Sandstone rests unconformably on Moinian and is followed by the Upper, which contains basic volcanics (Island of Hoy).

Mesozoic strata occur on both coasts of the northern part of the area and also under the Tertiary lavas which outcrop in the S.W. of the region.

Structural events were:-

Lewisian	Folding (and metamorphism)
Caledonian (Pre-Devonian)	Folding and thrusting (and metamorphism)
Late Caledonian	Folding and faulting
Hercynian	Faulting
Tertiary	Faulting

Since the Lewisian rocks were involved in Caledonian movements their pre-Moinian structures have been largely obscured. In places, however, Lewisian structures similar to those of the foreland (2.3) are preserved. Occasionally a band of epidiorite or serpentine has been mapped cutting the foliation of acid gneisses as do the basic dykes in the Lewisian of the kratogen. Such dykes have been recognised, for example, in the Glenelg Lewisian inlier (Barber and May, 1976). From their character, and from radiometric dates, it would appear that much of the Lewisian of the "inliers" belongs to the Scourian cycle (Watson, 1975, 29; Harrison and Moorhouse, 1976).

The Moinian (and Lewisian, where present) rocks have been affected by at least three major tectonic events (four have been recognised in some districts) which are Caledonian in the widest sense. Evidence of the more precise age for these events is discussed in section (3.4) as it is necessary to take into account the Moinian and Dalradian of the Grampian Highlands.

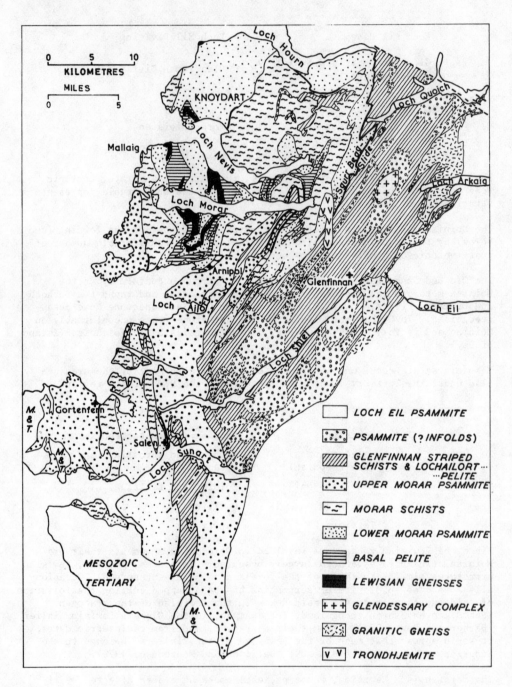

Fig. 3.3 GEOLOGICAL SKETCH MAP OF WESTERN INVERNESS-SHIRE, SCOTLAND (after Johnstone *et al.* 1969)

The earliest major tectonic event (F1) which affected the Moine rocks resulted in the interleaving, by isoclinal folding or thrusting, of Lewisian and Moine strata (Ramsay, 1958). Such interleaving has occurred in several parts of the Northern Highlands. These rocks are known to have been involved also in later folding during two major episodes (F2 and F3). In areas where Lewisian rocks are not present, F1 major or minor folds are not commonly clearly discernible, but sufficient evidence is generally available to show that the major structures are produced by the local F2 and F3 deformations. Normally the migmatization accompanying the Moinian metamorphism spans the period of the F2 folding (Dalziel and Johnson, 1963). F4, E.-W. folds have been recognised in some areas.

These broad conclusions may be illustrated from a few districts.

Near Glenelg, immediately E. of the Moine Thrust, Sutton and Watson (1958) recognised three fold-phases prior to the formation (considered to be Lower Devonian) of the Border Thrust-zone, namely isoclinal folding (F1) leading to the interleaving of Lewisian and Moinian, folding on easterly-dipping axial planes (F2) and folding under more rigid conditions on axial planes striking N.E. (F3).

Further S. the three highest groups of the Morar succession (see above) were first established by Richey and Kennedy (1939) in the Morar Antiform. At least three major deformations are now recognised (Poole and Spring, 1974). The earliest folds resulted in the isoclinal interleaving of Lewisian and Moinian (the Lewisian in part referred by Richey and Kennedy as sub-Moinian). The second phase produced a major westerly closing recumbent antiform, the Knoydart Fold, underlain by the Knoydart Slide. The open Morar Antiform belongs to the third phase and is flanked to the E. by the Ben Sgriol Synform. All the Moinian involved belongs to the Morar succession, which to the E. is separated from the Glenfinnan succession by the Sguir Beag Slide (Tanner, 1970), traceable for over 20 km S. from the head of Loch Hourn. This slide has brought up Lewisian in the Scardroy area.

The tectonics of the Glenfinnan division are complex and are characterised by a general steep deposition of the metasediments in long, attenuated flexures (probably F2), folded by younger (F3) structures which may be coaxial with F2 or may be oblique.

Further E. the Loch Eil division consists essentially of one formation, the Loch Eil Psammite, but there are infolds of Glenfinnan rocks. The Loch Eil division has been termed the "Flat Belt" but there are steeply-inclined zones, suggesting a large-scale basin and dome regime.

In Kintail Clifford (1957) postulated the formation of a Kintail Nappe (with a slice of Lewisian wedged between overlying Moinian and the underlying thrust) followed by folding on N.E. axes, and also the development of W.N.W. cross-folds. Further E. in the Fannich area Sutton and Watson (1954) have recognised N.-S. folds overturned towards the W., contemporaneous N.W. flexures, affecting a Moinian metasedimentary succession, complicated by what were later accepted as Lewisian tectonic wedges.

The Carn Chuinneag-Inchbae augen-gneiss in Ross-shire pre-dates the main Moinian fold-phases and has an aureole of hornfelsed sediments which have largely escaped the regional metamorphism. In the sediments N.W. folds are

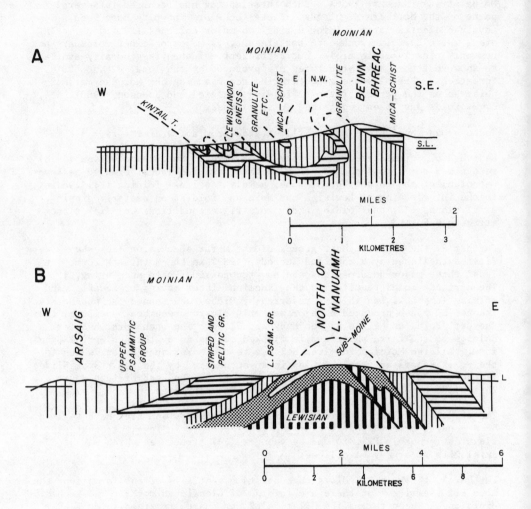

Fig. 3.4 SECTIONS OF NORTHERN SCOTTISH HIGHLANDS
    A, Section near Beinn Bhreac, Kintail, Ross-shire (after Clifford 1957)
    B, Section of southern end of Morar district showing the Morar Anticline (after Kennedy 1955).

preserved which are held by some to represent a pre-intrusion fold-phase and by others to be contemporaneous with intrusion.

In Sutherland Moinian stratigraphy and structure are difficult to establish because of involvement with major migmatite complexes. Long, narrow bodies of Lewisian-like rocks in the far N. may have been interleaved with Moinian by early folds.

The Moinian metamorphism ranges up to the sillimanite grade or granulite facies and is clearly polyphase, extending, according to Dalziel and Brown (1965), from before the F2 folds into the F3 phase. Retrograde metamorphism is evident in places. The sillimanite grade is largely associated with a belt of migmatization which extends through the Northern Highlands from the N. coast to Loch Linnhe.

Late Caledonian Structures

In Orkney, Middle Old Red Sandstone, resting unconformably on Moinian, is strongly folded, mainly on N.-S. axes although N.W. and N.E. folds also occur. This folding pre-dates the Upper Old Red Sandstone which is almost flat. In Caithness the much faulted Sarclet Pericline has a N.W. major axis; it is flanked to the N. by the Ackergill Syncline and to the S. by the Latheron Syncline. The Dunnet Head Sandstone belonging to the Upper Division is faulted against Middle Old Red Sandstone. The Old Red Sandstone of Orkney and Caithness is also cut by a number of faults, some of which are reversed. Some faults have a post-Upper Old Red Sandstone component. The northerly-striking Brims-Risa Fault of Orkney may be continued in Caithness by the Brough Fault which brings the Middle Old Red Sandstone over the Upper Old Red Sandstone of Dunnet Head to the W., at about 45°.

The Middle Old Red Sandstone of the Dornoch and Cromarty Firths lies in a major north-easterly syncline with subsidiary folds on the same axes. A basement (probably Lower Old Red Sandstone) is unconformably covered by the fossiliferous Middle Old Red Sandstone and, N. of Tain, is thrust over Moinian by pre-Middle Old Red Sandstone movements (Westoll, 1964).

Hercynian and Tertiary Movements

Hercynian and Tertiary faults are common in the Northern Highlands. It is, however, sometimes difficult to distinguish the age of the movements, some of which, in fact, may be Permian, to judge from relationships to camptonite dykes probably of that age.

The most powerful faults are on N.E. to N.N.E. and N.W. to N.N.W. lines. It is likely that the major displacements were Hercynian with sinistral transcurrent components for the N.E. faults and dextral for the N.W. fractures, e.g. the Loch Maree Fault which was also active in Precambrian times.

Some faults have important Mesozoic or Tertiary components. The N.E. Helmsdale Fault lets down a coastal strip of Trias/Upper Jurassic strata; it may continue S.W. through Dingwall and Strath Glass. The last movements in the Helmsdale district were post-Kimmeridge; Bailey and Weir (1933) have shown that it was an active submarine fault-scarp in Kimmeridge times from which were derived huge angular blocks of Old Red Sandstone now interbedded with normal marine Kimmeridge sediments.

North and S. of the mouth of the Cromarty Firth, Jurassic strata are let down against Middle Old Red Sandstone by N.E. faults which are close to the undersea prolongation of the Great Glen Fault.

In the S.W. of the region N. and N.W. faults (Macgregor, 1967) displace strata up to the Tertiary lavas. One of the most striking faults, scenically, is the N.-S. Inninmore Fault on the N. side of the Sound of Mull, which has a downthrow to the W. of at least 1000 ft (305 m), bringing down a discontinuous succession ranging from Coal Measures to Tertiary lavas against Moinian.

3.4   The Grampian Highlands (Scotland and Ireland) (Figs. 3.2, 3.5, 3.6, 3.7, 3.8, 3.9 and 3.10)

Apart from the N.E. the Grampian Highlands are noted for their rugged mountain scenery. Ben Nevis (4406 ft; 1345 m), within a Devonian ring-complex, is the highest point in the British Isles, and there are numerous other mountains, mostly of metasediments, rising to above 3000 ft (915 m). Structurally, the Grampian Highlands extend into Ireland where the mountains rise to nearly 2500 ft (760 m) in the N.W.

Exposures are excellent on the mountain ridges and on the sides of the deeply-glaciated valleys. Glacial drift, however, hides much of the solid geology in N.E. Scotland and in Ireland, apart from the N.W. The deeply-indented coast provides long rock-sections and in W. Ireland cliffs drop over 1500 ft (460 m) to the Atlantic.

Some excursions are described by Bassett (1958), Pitcher and Cheesman (1954) and Read (1960).

The Great Glen Fault (3.3) forms the N.W. boundary. In the S.E. the Highland Boundary Fracture-Zone extends for 380 miles (608 km) from Stonehaven on the North Sea to Clare Island off the Atlantic coast of Ireland. This is a complex zone, often consisting of several fractures, along which major movements, not all affecting the total length, have taken place from Arenig to Hercynian times. In Scotland, where it is often marked by ophiolites or by carbonated derivatives, it is almost continuously exposed. In Ireland considerable sections are hidden by Tertiary lavas in the E. and by Dinantian in the centre. Towards the W. its course is more doubtful. A fault along the S. shore of Clew Bay is regarded by some authors as the Boundary Fault. However, another fault-zone, not far to the S., separating Dalradian to the N. from Silurian to the S., contains serpentine and seems more likely to be the main Fracture-Zone. The offset of the fault-zone from the undoubted Highland Boundary on Clare Island (also with serpentine) could be due to the N.W. Doo Loch faults (3.5) or their _en echelon_ equivalent under the sea. It may be that on approaching the continental margin the Fracture-Zone not only becomes westerly, but splits up.

The formations present in the Grampian Highlands are:-

> Pliocene
> Tertiary volcanics
> Liassic
> Triassic)
> Permian) New Red Sandstone
> Carboniferous

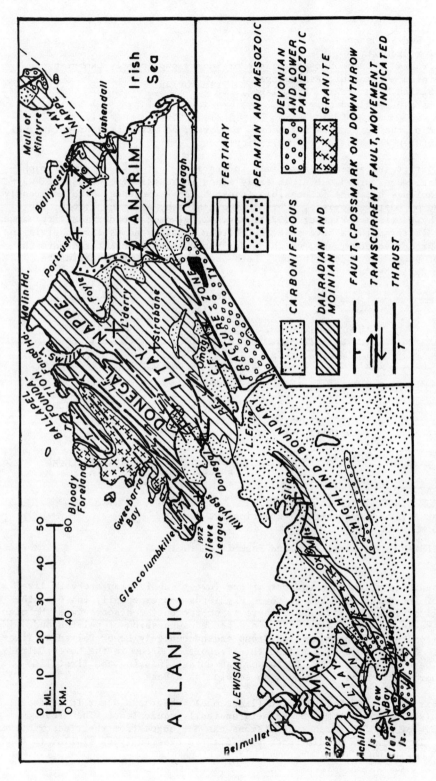

Fig. 3.5 STRUCTURAL MAP OF THE IRISH "GRAMPIAN HIGHLANDS"
C.F. = Cool Fault; L.F. = Leannan Fault; P.F. = Pettigo Fault;
T.P. = Tow Fault; Ty. = Tyrone Inlier.

Old Red Sandstone
Lower Ordovician
Dalradian Metamorphic Assemblage or Supergroup (partly Cambrian)
Moinian Metamorphic Assemblage) Precambrian
Lewisian Metamorphic Assemblage)
   (Mayo, Ireland only)

The base of the Moinian is not seen; moreover, no structural or metamorphic break between the Moinian and Dalradian occurs.

Many district successions, sometimes interpreted in the wrong stratigraphical order, were worked out at an early stage in research. The use of "way-up" techniques has made it possible to establish a sequence which is broadly accepted by most geologists. Correlation tables by Anderson (1965), Table I, and by Harris and Pitcher (1975), Fig. 12, give most older locality names. The two latter authors have selected certain locality names for subdivisions of the broad sequence, but, on the other hand, lithological terms have the merit of providing a description. Both are, therefore, set out in the summary table below:-

		Sub-groups and (groups)		
CAMBRIAN	(U. Psammitic Gr. (U. Pelitic and (Calcareous Gr.	(Southern Highland)	)))	UPPER
DALRADIAN	(L. Psammitic Gr. (L. Pelitic and (Calcareous Gr. (Carbonaceous Gr. (Quartzitic Gr. (	Crinan)) Easdale) (Argyll)) Islay)	)))))	MIDDLE
	(Calcareous Gr.	Blair Atholl) Ballachulish) (Appin)	))	LOWER
TRANSITION FROM DALRADIAN TO MOINIAN	Pelitic and Quartzitic Transition Gr.			
MOINIAN	Central Highland Psammitic Gr.			

The granulites which make up most of the lowest subdivision are very like those of the Moinian of the Northern Highlands and especially those of the Loch Eil division (3.3). Where there is no tectonic contact between Moinian and Dalradian there is a variable interbedded succession of pelites and quartzites between the Psammitic Group and undoubtedly Lower Dalradian limestones. Some authors have placed this Transition Group in the Lower Dalradian. There seems, however, to be some advantage in using the first limestone horizon in this continuous succession as a base.

The Blair Atholl and Ballachulish (limestone) sub-groups occur in different nappes (see below) but may be stratigraphically equivalent. The broad lithology of the Dalradian subdivisions can be judged from their descriptive names. In Ireland a Middle/Upper Dalradian succession, very similar to that

of the S.W. Highlands of Scotland, is present. Further S.W. in Ireland the
Middle/Upper Dalradian divisions are not as clearly defined, but what has
been termed the Kilmacrenan succession is almost identical with the Lower
Dalradian of the Iltay Nappe in Scotland (see below). North-west of the
large Donegal Granite the Creeslough Succession resembles that of the
Ballachulish district which belongs to a different nappe, that of the
Ballapel Foundation.

An important horizon at the base of the Quartzitic Group is a glacogene with
large boulders, often of nordmarkite, with interbedded upper and lower
contacts, which has been traced from Aberdeenshire to the Atlantic coast of
Ireland.

Near Loch Awe in Scotland and Strabane in Ireland, basic lavas, some with
pillow-structure, occur in the Upper Dalradian. Metadolerites are associated with these and, in fact, occur throughout much of the Moinian/Dalradian
outcrop.

Dating of the fold and metamorphic phases is discussed below. Stratigraphically, the Moinian and part of the Dalradian are late Upper Proterozoic.
The glacogene at the base of the Middle Dalradian (with limestone below and
quartzite above) is equivalent to the late Precambrian Varangian of Norway
(3.6). The Upper Dalradian, on the other hand, is Cambrian as pagetid
trilobites of late Lower Cambrian age occur near Callander, Perthshire, in
the Leny Limestone, which is part of the Upper Psammitic Group. As acri-
tarchs (Downie et al., 1971) of Lower Cambrian type occur down to the Loch
Tay/Tayvallich limestone horizon at the base of the Upper Dalradian, the
Precambrian-Cambrian boundary may be about here.

Narrow strips of basic lavas, black shales, cherts and other sediments occur
along the Highland Border, mostly separated from Upper Dalradian by faults
but in places, notably Arran, following on without stratigraphical or metamorphic break. Fossils, some doubtfully diagnostic, indicated Upper
Cambrian/Lower Ordovician ages.

Many late Silurian/Lower Devonian, post-tectonic, granitic intrusions are
present, some in ring-complexes (for structure see below).

Considerable outcrops of Lower Old Red Sandstone sediments and volcanics
unconformably overlie the metamorphic rocks, notably in Lorne (near Oban) in
Aberdeenshire and immediately N. of the Highland Border. Middle and Upper
Old Red Sandstone conglomerates and sandstones cover a large area round
Inverness and S. of the Moray Firth. Upper Old Red Sandstone sediments
directly overlie the Dalradian in the Loch Lomond/Firth of Clyde district,
where they are overlain by a small outcrop of Lower Carboniferous strata. A
patch of Upper Carboniferous occurs along the Pass of Brander Fault E. of
Oban.

The "Grampian Highlands" in Ireland differ from the corresponding region in
Scotland in the presence of extensive outcrops of Lower Carboniferous strata,
including thick limestones. Throughout much of the region sedimentation
seems to have started in the Visean, but in the Omagh district a considerable
thickness of sandstones and sandy limestones below C_2S_1 beds are Tournaisian.
In the extreme N.E., at Ballycastle, there is a Lower Carboniferous succession, including basalt lavas, similar to that of the Midland Valley of Scotland, followed by sandstones, shales and coals belonging to the Upper

Carboniferous.

In N.E. Scotland Triassic strata, with reptiles, unconformably overlie the Old Red Sandstone. In the S.W., in Kintyre, Permian red sandstones occur. In N.E. Ireland Triassic red conglomerates, sandstones and marls occur round the rim of the Antrim Tertiary basalt lavas. These are followed in places by Rhaetic and Lower Liassic shales and thin limestones, followed by Upper Cretaceous white limestone, a hard chalk. In both W. Scotland and in Ireland there are Tertiary dyke-swarms (9.2).

Structural events were as follows:-

    Caledonian (Pre-Lower Devonian) - folding (and metamorphism)

    Structures connected with Lower Devonian intrusions

    Late Caledonian (Pre-Upper Devonian)

    Probably Hercynian

    Probably Tertiary

The polyphase folding and polymetamorphism started (in some regions at any rate) in Sardic (Upper Cambrian) times and continued to the Silurian (for discussion see below).

It is generally accepted that most, if not all, of the Dalradian was affected by three major fold-phases (Johnson, 1969), although some authors have postulated more; for example Harris, Bradbury and McGonigal (1976) recognise four phases in the Tay Nappe and Roberts (1974) no less than eight in the S.W. Scottish Highlands.

The nappe hypothesis for the Grampian Highlands originally put forward by Bailey (1922) has been greatly modified. Nevertheless, it is accepted that a large part of the "Grampians", both in Scotland and Ireland, is made up of the generally north-easterly striking Iltay Nappe-Complex. This structurally overlies the Ballapel Foundation seen mainly to the N.W. of the Iltay Nappe both in Scotland and Ireland. A Banff Nappe has been postulated in the N.E. Grampians but some authors regard this as part of the Iltay Nappe.

The Iltay Nappe-Complex is largely characterised by N.E. trending recumbent folds (F1) accompanied by sliding. These close towards the S.E. in the S. of the Nappe-Complex, towards the N.W. in the N., the two parts being separated by a belt of steeply inclined structures (Harris, 1963). The F1 folds are modified by fairly open cross-folds (F2) and folds (F3), trending N.E. From near Kirkmichael to Kintyre the lower limb of a major recumbent anticline forms the "Loch Tay Inversion" affecting mainly Upper Dalradian metasediments. The upper limb is preserved in a notable F3 fold, the Loch Awe Syncline, where the metasediments (Allison, 1940) and volcanics occur in normal order. Still later (F4) minor folding and buckle folding can also be recognised in some districts. The Loch Tay Inversion may continue to the Highland Border and be flexured downwards close to the Highland Boundary Fracture-Zone (Shackleton, 1958). To the N.W. of the steep belt the overfolds, directed N.W., are shorter and are separated by the Iltay Boundary Slide from the folds of the Ballapel Foundation (see below).

South-easterly plunging structures are a marked feature near the Cairngorm

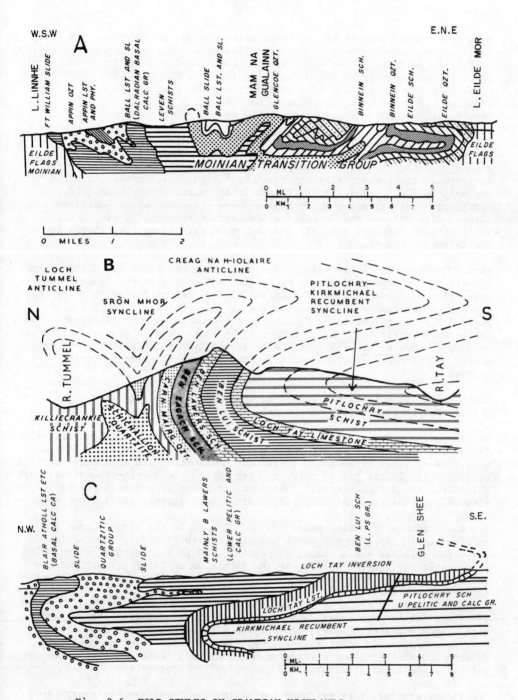

Fig. 3.6 FOLD-STYLES IN GRAMPIAN HIGHLANDS
  A. Lochaber (North West) (after Bailey)
  B. Perthshire (centre) (after Harris)
  C. Eastern Perthshire (South-East) (after Bailey)

Mountains where there is a structural swing southwards. Further N.E. the folds may be autochthonous or parautochthonous and may have been relatively held back, possibly by the resistance of the large and numerous pre-tectonic basic intrusions which occur here, while those to the S.W. were driven more powerfully south-eastwards accompanied by more intense overfolding and by sliding. It is perhaps significant that the Iltay Boundary Slide has not been recognised N.E. of the southward swing.

The rocks of the Ballapel Foundation differ in facies from those of the Iltay Nappe. Stratigraphic discordances were accounted for by Bailey by slides dividing the unit into nappes, but they may be largely due to facies variation; the lowest Dalradian rocks rest along a sedimentary contact on the Moinian (Voll, 1964, 590). This Moinian/Dalradian continues N.E., with stratigraphical variation which includes the thinning out of quartzites, through Glen Roy and the Monadliath Mountains (Anderson, 1956; Piasecki, 1975) to join with the probably autochthonous structures (see above) round the Cairngorm Granite.

In N.W. Ireland the Iltay Nappe, including the Lower Dalradian Kilmacrenan Succession, is separated partly by a slide but mainly by the Donegal Granite from Ballapel Foundation, characterised by a different facies, that of the Creeslough Succession; this succession includes Moinian. Further S. in Ireland Moinian occurs in an F3 anticline around Lough Derg (Anderson, 1948), in the Ox Mountains and in North Mayo. The Lough Derg Moinian psammites are held by Pitcher et al. (1971) to be separated from pelites further N. by a thrust, but the evidence is doubtful; the thrust, if present, may be further N.

Gneisses in North Mayo, characterised by pre-Moinian structure and metamorphism, are correlated with the Lewisian (Sutton, 1972).

Metamorphic zoning on the basis of certain index minerals was established by Barrow (1893) in the S.E. Grampians and has served as a model for similar work in many other metamorphic terrains. Andalusite- and cordierite-bearing schists occur in the N.E.; this has been termed the Buchan-type of metamorphism from the Buchan district, Aberdeenshire, and is probably due to elevation of the regional temperature at a comparatively shallow depth.

Two early phases of metamorphism (M1 and M2), up to the garnet grade (or amphibolite facies), appear to be associated with F1 and F2. Dolerite intrusion followed and a phase, or phases, of high temperature metamorphism in a static environment leading to the formation of later garnet, kyanite and sillimanite in both Scotland (S.E. Grampians, S. Inverness-shire) and Ireland (Lough Derg).

The three (or in some cases four) fold-phases represent distinct and separable orogenic pulses. Where, however, more phases have been described it must be asked, as in other regions, what is the nature of true polyphase folding and whether minor structures of relatively limited distribution may not be due to local structural and geological inhomogeneity.

The rather contradictory evidence for dating the folds (and metamorphisms) affecting the Moinian/Dalradian successions of the Northern Highlands (3.3) and of the Grampian Highlands (including Ireland) can be discussed only briefly.

CALEDONIAN - METAMORPHIC

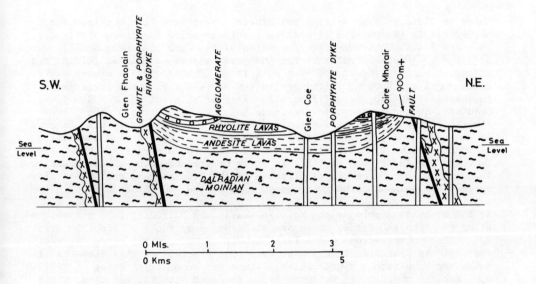

Fig. 3.7 SECTION OF LOWER DEVONIAN CAULDRON - SUBSIDENCE OF GLEN COE

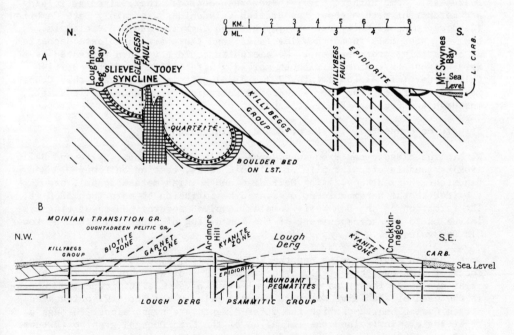

Fig. 3.8 SECTIONS OF IRISH "GRAMPIAN HIGHLANDS"
  A. Across Slieve League Peninsula, Donegal
  B. Across Loch Derg, Donegal.

Dates of 740 m.y. (van Breeman and others, 1974) have been obtained for
pegmatites in the Moinian of Western Inverness-shire (Northern Highlands),
and some folding preceeded the intrusion of the Carn Chuinneag Granite dated
at 560 m.y. Precambrian dates for the early phases of Moinian folding and
metamorphism has, therefore, been postulated (Powell, 1974). These early
dates conflict, however, with many in the 420-450 m.y. range (regarded by
Moorbath, 1975, 110, as uplift-cooling ages of an event about 500 m.y. old).
Moreover, if the earliest fold-episode is coeval with that of the Moine
thrust-belt, then it is post-Arenig. The Moinian sedimentary pile is over
10 km thick, and pegmatite intrusion could have gone on in its base while
sedimentation continued; alternatively, it has been suggested that in the
Northern Highlands (and possibly in parts of Ireland, e.g. the Ox Mountains)
there is an "Old Moine", metamorphosed in the Precambrian, and a more widespread "New Moine", although no break between the two has been found.

In the Grampian Highlands the Moinian shares the Dalradian fold phases, and
along the Highland Border these are also shared by rocks of probable Arenig
age. These observations and a large number of radiometric dates suggests
that all the Dalradian fold-phases were post-Arenig; on this view the F1
folds may have been Mid-Ordovician. There are a number of dates of 390-440
m.y. for the progressive metamorphism associated with F2, and these would
lead to the conclusion that F3 and F4 were Silurian or even early Devonian.

In Connemara (3.5) almost unaltered Arenig is in contact with metamorphosed
Dalradian. Some authors have, therefore, concluded that Dalradian folding
and metamorphism throughout the British Caledonides was mainly Late Cambrian
(Sardic). Unconformity between the British Cambrian and Ordovician is,
however, seen only in North Wales (4.2), and there are no molasse deposits
of a possible Sardic orogeny. Metamorphism could have been initiated as a
Sardic event in Connemara without markedly affecting most of the British
Caledonides (cf. Alpine events, 7.1 - 7.14). It is significant that Connemara is separated from the rest of the Dalradian by an ancient lineament,
the Highland Boundary Fracture-Zone.

Structures Connected with Lower Devonian Intrusions

Within the classic cauldron subsidence of Glen Coe (Clough, Maufe and Bailey,
1909) a roughly circular pile of volcanic rocks, resting on Lower Old Red
Sandstone sediments overlying Dalradian and Moinian metasediments, has
subsided to the accompaniment of ring-dyke intrusion by more than 3000 ft.
The formation of the Ben Nevis Complex involved several phases of ring-dyke
intrusion to the accompaniment of the foundering of a central block of Lower
Old Red volcanics.

Both the Ben Nevis Complex and the Etive Complex, S. of Glen Coe, are the
focus of great north-easterly dyke-swarms. The dilation of the crust at
right angles to these was a structural event of some significance, amounting
in the case of the Etive swarm to about a mile. Near the N.W. orientated
Strath Ossian quartz-diorite the metasediments are swung almost through a
right angle. In Ireland the intrusion of the Main Donegal granite (Pitcher
and Read, 1960) has caused marked folding of its envelope.

Post-Lower Devonian/Pre-Upper Devonian Structures

The main movement along much of the Highland Boundary Fracture-Zone was of
post-Lower/pre-Upper Old Red Sandstone age. The Lower Old Red Sandstone

A

B

Fig. 3.9 A. Folded metadolerite boudin in metapsammite of Shetland Metamorphic Series, West side of Lunna Ness, Shetland.
B. Bedding and cleavage in shale and greywacke of Barents Sea Series.
Kongsfjord, Varanger Halvoya, Norway

outliers N. of the fracture-zone (Allan, 1928, 1940) are folded along N.E. axes and are cut by normal or possibly reversed faults, parallel to, and at about 30° to, the Highland Boundary Fault. These folds and faults are likely to have been of Middle Old Red Sandstone age and, if this is so, their presence in the Grampian Highlands would support the view that the last folds affecting the metamorphic rocks may be as late as Middle Devonian. The Middle Old Red Sandstone, S. of the Moray Firth, is fairly gently folded along N.E. to E.N.E. axes which predate the unconformably overlying Upper Old Red Sandstone.

Hercynian (sensu lato) and Tertiary Structures

An important structural feature of the region (and of the Northern Highlands, 3.3) is the presence of many N.E. to N.N.E. faults; some of these have probably had a long history extending back to the Lower Palaeozoic. Yet important components of movement are later. Some are certainly later than the post-tectonic Caledonian granites and some, in the N.E. Grampians, displace Middle Old Red Sandstone, and others, near the Firth of Clyde, the Upper. In Ireland some show marked post-Visean displacements. For a number of these faults powerful sinistral transcurrent movement can be demonstrated, e.g. the Ericht-Laidon Fault (4½ miles, 7 km), the Loch Tay Fault (4 miles, 6 km) and faults cutting the Barnemoore granite in Donegal (3 miles, 5 km). For the powerful Leannan Fault, further N.W. in Donegal (Pitcher et al., 1964), correlation with the Great Glen Fault has been suggested, but it is more likely a splay fault to the latter.

Much of the faulting was probably Hercynian (sensu lato) in response to N.-S. compression. The predominant N.E. grain would favour development in this direction. Complimentary N.W. fractures are, in fact, much less common, but there are a number of minor N.W. faults.

In Ireland the Lower Carboniferous rocks resting on the metasediments are thrown into gentle anticlines and synclines. Many of these folds have a N.E. trend, and their development as Hercynian folds is due to the influence of the underlying Caledonian structure. Folds in the Clew Bay area run E.-W., again in accord with the underlying structure.

Tertiary movements can be demonstrated in N.E. Ireland. They include the accentuation of the anticline along the Highland Border and the development of downfolds to N.W. and S.E. Faulting took place along the N.W. and S.E. margins of the Highland Border ridge and also at right angles. The scale of the deformation may be judged from the fact that combined folding and faulting has led to differences in level of the Chalk of over 3000 ft.

3.5   Connemara and its Flanks (Ireland) (Fig. 3.10)

This mountainous region, rising to over 2600 ft (730 m) and deeply-indented along the Atlantic coast, is bordered by Clew Bay and Galway Bay. To the E. its Lower Palaeozoic and older rocks are unconformably overlain by the Lower Carboniferous of Central Ireland (6.7).

The formations present are:-

>   Silurian
>   Ordovician
>   Dalradian

Although the sequence is interrupted by slides, a Lower/Middle Dalradian sequence has been recognised. This includes, as in Donegal and Scotland, a boulder bed of glacogene type with a carbonate horizon (containing the well-known Connemara Marble) below and a quartzite above.

The Ordovician occupies much of the wide South Mayo Trough. Arenig clastics at the base are interbedded with spilites and basic tuffs. Slates, sandstones and grits follow, succeeded by greywackes and conglomerates ranging up to the Caradoc and containing ignimbrites (McKerrow and Campbell, 1960). To the S. the Ordovician is unconformably overlain by an Upper Llandovery conglomerate followed by a varied clastic succession up to the Wenlock. On the S. margin of this Silurian succession, however, Wenlock basal conglomerate rests directly on Dalradian.

Between the Ordovician and the serpentine-bearing fault to the N., regarded (see above) as the main branch of the Highland Boundary Zone, a Silurian succession of a different character occurs (Anderson, 1960; Bickle et al., 1972). The Croagh Patrick Quartzite with a basal conglomerate is followed by pelites and calcareous pelites succeeded by psammites. In spite of epizonal metamorphism, distorted fossils are preserved and prove a Wenlock age.

In places on the S. margin the Wenlock rests unconformably on a thin Llandovery succession, and to the W. on Clare island it overlies Dalradian and possibly older rocks (Philips, 1973, 1974).

In the S. of the region the large Galway Granite, dated at 365 m.y., was intruded post-tectonically. On its S.W. margin hornfelsed basic pillow-lavas, tuffs and sediments with Dalradian clasts (the South Connemara Series) are assigned to the Ordovician.

Structural events were:-

Caledonian	(Late Cambrian (- Late Ordovician	- folding and metamorphism
	(Post-Wenlock (- Pre-Carboniferous	- folding and metamorphism
Hercynian and Tertiary	Post-Carboniferous	- mainly faulting

Four fold-phases have been recognised in the Connemara Dalradian (Badley, 1976; Yardley, 1976), and Barrovian metamorphism ranges up to the sillimanite grade and migmatites.

A D1 deformation, evidence for which rests mainly on fabric studies, was followed by upright folds (D2) which were later flattened. Northward-facing nappes (D3) were then formed. Open D4 folds developed after the rocks had largely cooled and include the most obvious structures such as the Connemara Antiform. The time-relationship of the D4 folds to the overthrusting of high-grade rocks over low-grade schists to the S. is not clear.

As Arenig strata with metamorphic clasts come against the schists it has been claimed that all the metamorphism was pre-Arenig. However, as the contacts are invariably large faults this conclusion is by no means certain.

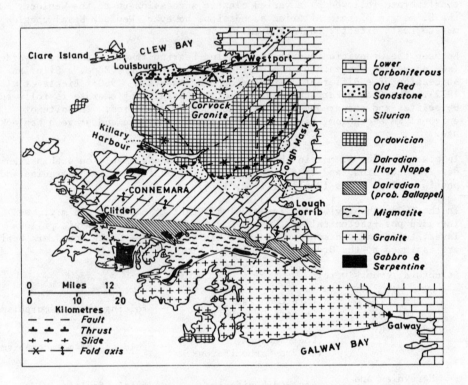

Fig. 3.10 STRUCTURAL MAP OF CONNEMARA AND ITS FLANKS
C.P. = Croagh Patrick

The clasts could have come from older metamorphics or they could have been derived from a part of the Connemara pile which had undergone one of the earlier metamorphisms and had been uplifted at an initial stage of the formation of the Connemara Cordillera and South Mayo Trough. From a date of 510 m.y., obtained by Pidgeon (1969) for metamorphism in the Connemara schists, and from other evidence, the early folding and metamorphism would appear to be due to late Cambrian (Sardic) events, but folding and metamorphism continued into the Ordovician; Phillips (1973) suggests a mid-Ordovician age. Metamorphism of this age could have faded out rapidly northwards and not affected the Ordovician, just as the undoubted post-Wenlock metamorphism S. of Clew Bay (see below) diminishes rapidly southwards. D4 in the Dalradian could be contemporaneous with part of the folding in the Ordovician.

In the Ordovician the main structure is the Mweelrea Syncline, which is flanked to the N. and S. by anticlinal zones. There is, however, some doubt as to the relative importance of pre-Silurian and post-Silurian (see below) folding in determining the final form of these folds (McKerrow and Campbell, 1960, 44).

Structures of Post-Wenlock and Pre-Carboniferous Age

There is evidence of intra-Silurian movement, for Lower Wenlock conglomerates rest unconformably in places on Upper Llandovery. However, the main folding is post-Wenlock. Whether it was late-Silurian or Devonian remains in doubt. The folds trend E.-W. in most of the region but in the E. swing into the N.E. Caledonian strike.

To the N. lies the Croagh Patrick Syncline overturned towards the S. so that the Silurian succession on its northern limb is inverted at angles as low as $30°$ and folded into an antiform (Anderson, 1960). The folding was accompanied by epizonal regional metamorphism; this is strongest in the N. and fades out (apart from cleavage) in the Ordovician strata S. of the Silurian. The main folding and metamorphism was followed by N.W. cross-folding. The post-tectonic Corvock Granite has imposed strong thermal metamorphism on the regional.

The Silurian strata sandwiched between the Ordovician and the Connemara schists form several wide synclines, with steep northerly limbs and intervening narrow anticlines.

The Salrock Fault is an important fracture which belongs to the post-Wenlock compressional phase. This has an easterly course roughly following Killary Harbour and is here a low-angled overthrust with Ordovician pushed upwards on the N. side. Further E. it becomes a high-angled overthrust and, like the folds, swings N.E. This fault has been regarded as the continuation of the Southern Uplands Fault.

Post-Carboniferous Structures

A number of N.E. and N.W. faults, on many of which transcurrent displacements can be demonstrated, cut the region. Probably both Hercynian and Tertiary fracturing has occurred.

3.6 Shetland (Figs. 3.9 and 3.11)

The Shetland Islands are the highest parts of a mostly submerged, irregular Palaeozoic and older block separating two large submarine sedimentary basins (8.13). The deeply-indented rocky coast is fringed in places by cliffs over 500 ft (152 m) high. Inland, hills rise to 1475 ft (450 m) above peaty moorlands. The islands lie about 103 miles (165 km) N.E. of the Scottish mainland and about 210 miles (336 km) from the nearest point in Norway.

The formations present are:-

>   Upper Old Red Sandstone
>   Middle Old Red Sandstone
>   Lower Old Red Sandstone
>   Shetland Metamorphic Rocks, including units of
>   Dalradian, Moinian and (?) Lewisian ages.

There are marked contrasts in the geology E. and W. respectively of the Walls Boundary Fault, which is probably the continuation of the Great Glen Fault (3.3).

Acid and hornblendic gneisses of Lewisian aspect form the N.W. corner of the Mainland. These could be part of the Caledonian foreland or else a Northern Highland-type inlier (3.3). The rest of the metamorphic rocks W. of the Walls Boundary Thrust consist of impure quartzite, hornblendic gneiss and muscovite-schist followed by green schists and calcareous rocks. These metasediments are unlike those E. of the Fault. Although radiometric dates give a Caledonian age it has not been possible to equate them with any members of the Moinian/Dalradian sequence.

The metamorphics of the Mainland E. of the Walls Boundary Fault form a long succession starting with psammites (granulites) like those of the Moinian; similar rocks make up Yell. On the Mainland the psammites are followed by quartzites, pelites, metalimestones, grits and altered spilitic lavas which can be tentatively correlated with the Dalradian of the Grampians (3.4). There are two belts of migmatites, granites and pegmatites. Along part of the E. coast of the Mainland gneisses, semipelites and gritty limestones, tectonically separated from the metasediments to the W., are termed the Quarff Succession.

In Unst and Fetlar serpentinites, metagabbros, etc., are involved in a Nappe Pile. Post-tectonic Caledonian granites occur in both the E. and W. Mainland.

The Walls Sandstone, of Lower/Middle Old Red Sandstone age, outcrops in the W. and contains basalt, andesite and rhyolite lavas as well as ashes. Upper and high Middle Old Red Sandstone strata occur near Lerwick.

Some excursions are briefly described by Mykura (1976, 138-143).

The main structural events were:-

>   Pre-Caledonian
>   Caledonian - pre-Devonian
>   Late or Post-Devonian

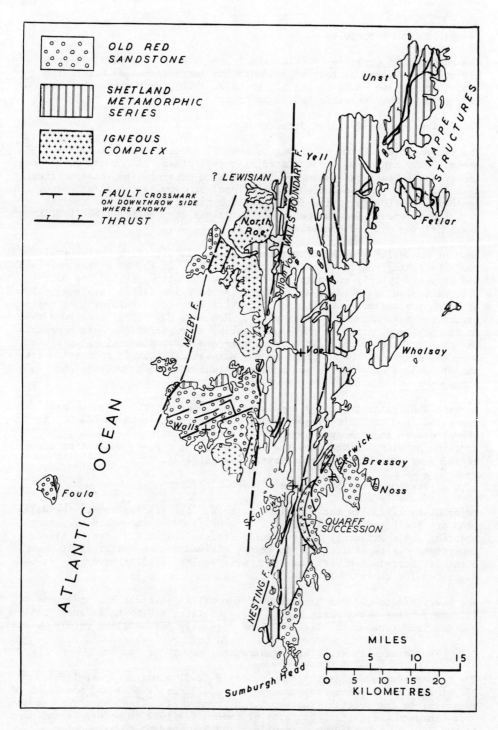

Fig. 3.11 STRUCTURAL MAP OF SHETLAND

Pre-Caledonian Structures

The gneisses of probable Lewisian age in the N.W. of the Mainland were affected by several periods of metamorphism and deformation before the deposition of the rocks to the E. (Pringle, 1970), from which they are separated by the Wester Keolka Shear Zone.

Caledonian - Pre-Devonian Structures

The metamorphics to the E. belong to several geographically separated groups, only tentatively correlated with each other. All these rocks have undergone at least two phases of folding with metamorphism, radiometrically dated as Caledonian, accompanying or closely following the first phase. The southerly-dipping foliation in the Walls Peninsula must be due to late Devonian movements (see below) as it parallels the southerly bedding dip of the overlying Old Red Sandstone (Mykura, 1976, 23).

East of the Walls Boundary Fault the tectonic fabric of the metamorphic rocks (striking mostly N. by E. but bending N.E. in the N. of the Mainland) was produced by only one phase of intense deformation, termed the Main Deformation, of Caledonian age. According to Flinn (1967) this was followed by porphyroblast metamorphism and migmatisation, etc. Refolding of earlier structures has, moreover, taken place. However, May (1970) concluded that in the Scalloway district the migmatisation accompanied the main movements and that the latter post-dated an earlier and first phase of metamorphism. Near the middle of the E. coast of the Mainland the Quarff succession (see above) forms a nappe which was pushed westwards along a "melange" or schuppen-zone.

In Unst and Fetlar two major nappes of basic and ultrabasic rocks are separated from each other, and from the metamorphic rocks to the W., by thrust-planes and schuppen-zones. The zone separating the two nappes contains pebbles derived from the erosion of the nappes, suggesting nappe-emplacement at a high tectonic level (Mykura, 1976, 7).

Late or Post-Devonian Structures

Major N. by E. faults include, from E. to W., the Nesting Fault, the Walls Boundary Fault and the Melby Fault. The Walls Fault is marked by intense shearing and shattering. In the Walls Peninsula an E.N.E. set is also important. Large dextral displacement, claimed for the Walls Fault, complicates its correlation with the Great Glen Fault; similar movement has been suggested for other faults.

The Lower/Middle Old Red Sandstone of the Walls Peninsula has undergone intense folding along early E.N.E. axes and later N.N.E. to N. axes. The Old Red Sandstone of the East Mainland is gently folded along mainly N. axes.

3.7 The Caledonian Front in Scandinavia (Norway and Sweden) (Figs. 2.1, 3.12, 3.13 and 3.14)

The Caledonides override the Baltic Shield (2.1) along a thrust-front for 350 km from the Barents Sea, through the moorlands (vidda) and mountains of Finnmark, to the Swedish frontier E. of Tromso. The front then runs southwards through the border mountains of Sweden for 1000 km before re-entering Norway N.E. of Lake Mjosen. Here it turns S.W. and W. to extend for a further 500 km through the high mountains of south central Norway to the sea

S. of Bergen. For most of this 1850 km, the front is well-exposed, being hidden under glacial drift for only relatively short distances, for example near Lake Mjosen.

The broad structural succession is as follows:-

Allochthonous nappes	-	Caledonian (and probably also Precambrian) metamorphics
Parautochthonous nappes	-	Eocambrian (Vendian or Riphean) and Cambro-Silurian sediments
Autochthon	-	Cambrian/Ordovician on Precambrian metamorphic complexes

The Cambrian/Ordovician below the sole thrusts of the front is generally thin, due, in part, it is believed, to the bull-dozing action of the advancing nappes and in part to an originally eastward thinning sequence. Although the sedimentary nappes are referred to as parautochthonous there is evidence that some are far travelled; even greater distances (up to 550 km) are claimed for the metamorphic nappes. "Basement" rocks appear in culminations far to the W. of the front. Mylonite is extensively developed along some of the thrusts.

The structures between the autochthon and the eastern or southern edge of the metamorphic nappes are described in the present section by reference to three districts - Finnmark, Swedish Border Mountains and Finse.

Finnmark

In the extreme N. of Norway the N.W. margin of the Baltic Shield forms an autochthon, consisting of a Karelian "basement" (2.1) unconformably overlain by the Eocambrian to Lower Cambrian Dividal Group, consisting mainly of thinly-bedded quartzites and shales.

This is overridden by the Gaissa Nappe (Rosendahl, 1945), the lowest thrust unit of the Finnmark Caledonides, for which a minimum translation of 35 km towards the S.E. has been estimated (Gayer and Roberts, 1971), as shown by exposures on both sides of the Lakselva Valley.

The nappe forms a wide outcrop round the head of the Porsangerfjord; to the N.E., S.E. of the Laksefjord, it passes gradually into the supposedly autochthonous sediments of the Tanafjord Group (Føyn, 1967).

The Gaissa Nappe consists of the Porsangerfjord Group, comprising the Gaissa Sandstone Formation, an alternating sequence some 600 m thick of quartzites and interbedded shales, siltstones and sandstones, overlain by 170 m of white, stromatolitic dolomite of the Porsanger Dolomite Formation. Further E., towards Laksefjord, the Porsanger Dolomite is overlain unconformably by Varangian tillites and younger Eocambrian to Lower Cambrian sediments of the Vestertana Group.

The rocks of the nappe are not metamorphosed but have undergone three phases of deformation. The first phase formed buckle folds of various styles about approximately N.-S. axes with associated lithology-dependent cleavages. Late stage thrusting was a significant feature of this phase. The second

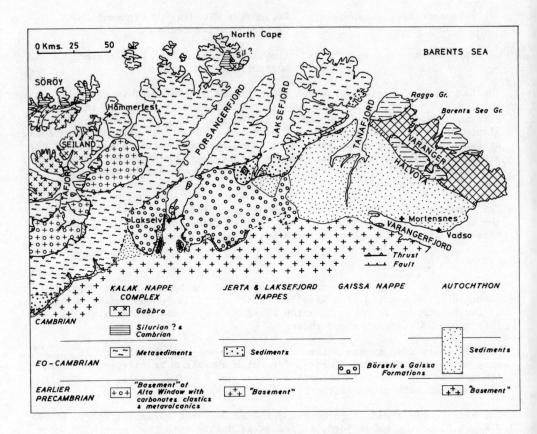

Fig. 3.12 STRUCTURAL MAP OF FINNMARK, NORWAY
(after Gayer and Roberts 1973)

phase of deformation produced localised crenulation of the D1 cleavage and more widespread buckle re-folds. The D3 deformation produced large-scale E.-W. warps and normal faulting.

To the N. and W. the Gaissa Nappe is overridden by the metamorphic allochthonous units of the Laksefjord and Kalak (Kolvik) Nappes (3.8). Spectacular sections of the Kolvik Thrust are seen above the shores of the inner Porsangerfjord where the dark metasediments rest on the white Porsanger Dolomite.

Swedish Border Mountains

Ostersund District

The structures of the E. margin of the Caledonides are excellently displayed on a readily accessible traverse from Ostersund towards the Norwegian frontier (Asklund, 1960; Thorslund and Jaanusson, 1960; Gee, 1975). The traverse, in fact, can be easily continued westwards to Trondheim (3.8).

In the Ostersund district the Precambrian "basement" crystallines are overlain by the Jamtland Supergroup. In the parautochthonous nappes this includes late Precambrian to Silurian sediments; in the foreland or autochthon only Cambrian/Ordovician strata occur. The autochthon, along with the parautochthonous nappes and the lowest of the allochthonous nappes (the Offerdal Nappe), is referred to an Eastern Complex in contrast to an overriding Western Complex of low to high-grade metamorphic nappes.

East of Ostersund the autochthon is made up mainly of Svecofennian (2.1) schists and gneisses intruded by granite dated at about 1785 m.y. These are overlain by the Jamtland Supergroup, consisting essentially of Cambrian black alum shales and Ordovician limestones. Immediately S.E. of Ostersund there is an overlap of beds up to the Caradoc onto the "basement".

A wide belt to the W., around Ostersund and the Storsjon lake, is occupied by a pile of thrust-slices, referred to as the Jamtlandian Nappes. The thrusts are at low angles and are themselves folded. The lowest, or sole, thrust is a decollement over the "basement". The nappes consist largely of Ordovician sediments with shales and greywackes in the W. giving way eastwards to limestones. The Ordovician sediments are overlain conformably by Silurian quartzites and limestones ranging up into a greywacke sequence containing clasts preserving a Ludlow fauna.

The Cambrian occurs beneath the Ordovician, and not far below there is a tillite correlated with the Varangian. A basal sequence of Riphean arkoses occurs in the higher nappes. Slices of "basement" occur at the base of the nappes; their relationship to the decollement is controversial.

North of the W. end of Storsjon the Jamtlandian Nappes are overridden by the allochthonous Offerdal Nappe along a thrust marked by thick (often tens of metres) mylonite. The Offerdal Nappe includes late Precambrian arkoses as well as Precambrian granites and acid volcanics.

Between Storsjon and the Norwegian frontier the Eastern Complex passes under the Sarv Nappe (Stromberg, 1955), the basal unit of the Western Complex. This nappe consists of shallow-water sediments penetrated by a tholeiitic

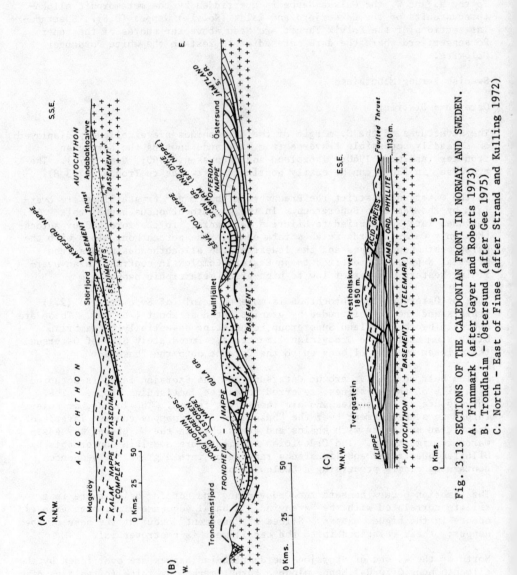

Fig. 3.13 SECTIONS OF THE CALEDONIAN FRONT IN NORWAY AND SWEDEN.
A. Finnmark (after Gayer and Roberts 1973)
B. Trondheim - Östersund (after Gee 1975)
C. North - East of Finse (after Strand and Kulling 1972)

dyke-swarm; thermal metamorphism by the latter has preserved sedimentary structures. If correlation with a unit of similar facies near the Norwegian Atlantic coast be accepted, the Sarv Nappe has moved E. over at least 200 km. The Nappe is overridden by the Sevi-Koli Nappe-Complex (3.8).

The Finse District

For 120 km W. of Lake Mjosen parautochthonous nappes of Eocambrian (Sparagmitian) override autochthonous Archean with a thin Cambrian cover. Sixty km E. by N. of Finse the parautochthonous nappes are overthrust by the S. edge of the huge Jotun Nappe which makes up much of south central Norway (3.8). The 35 km-long ridge of the Hallingskarvet, N.E. of Finse, and the ice-capped Hardangerjokelen to the S., are two large klippes, up to 22 km, S. of the main nappe front (Strand and Kulling, 1972, 46). Generally the thrust lies about 300 m above the Precambrian peneplain but there may be as little as 7 m of Cambrian. On the N. face of the Hardangerjokelen contorted, Cambrian, dark phyllites with sandy beds, overlying the Precambrian "basement", form the lower slopes. The Cambrian is overlain at the foot of the main cliff by mylonite, which passes up into acid and basic high-grade schists or gneisses which show through the ice-cap; these rocks are affected by at least two fold-phases.

From the Hardangerjokelen S.W. to the Boknfjord a number of crystalline massifs overlie the Cambrian phyllites. If these are klippes of the main Caledonian Nappes (although this has been disputed by some authors) they provide evidence of translation of at least 120 km.

3.8   The Caledonian Fold-Belt in Scandinavia (Norway and Sweden) (Figs. 3.15 and 3.16)

The Caledonian Fold-Belt extends from the Barents Sea to the Atlantic coast S. of Bergen. In south central Norway it forms the Jotunheim - the home of the giants - where the Glittertind, the highest part in Scandinavia, rises to 2470 m. The mountains on the Swedish/Norwegian border reach just over 1900 m, but in Finnmark the topography is lower, seldom exceeding 1000 m. Exposures are excellent in the mountain areas and on the spectacular rock-walls of the fjord-indented Atlantic coast.

Excursions to relatively accessible areas, both in the fold-belt and in the thrust-zone, are described by Gee, 1975; Holtedahl et al., 1960; Kvale, 1960 and Strand and Holmsen, 1960.

The geological succession is as follows:-

                Cretaceous
                Jurassic
                Middle Devonian
                Lower Devonian
                Silurian
                Ordovician
                Cambrian
                Eocambrian or Sparagmitian
                    (= Vendian or part of Riphean)
                Archean and Lower Proterozoic

The pre-Sparagmitian probably includes rocks formed during several of the

Fig. 3.14 A. Boudinage in granulite of Kalak Nappe Complex Veidnes, W. side of Porsanger Fjord, Finnmark, Norway
B. Hornblende-gneiss (mountain cliff) thrust over imbricated Valdres sparagmite.
Bitihorn, near Bygdin, South Central Norway

CALEDONIAN - METAMORPHIC

cycles which affected the foreland (2.1 and 2.2). Within the fold-belt they occur both as autochthonous "basement" and as tectonic enclaves.

The Sparagmitian consists of a succession of dominantly clastic sediments (Gr. *sparagma* = a fragment) which varies in thickness from several thousand metres within the fold-belt to a few hundred metres on its margins, where higher members overlap onto the "basement". The unaltered sediments of some districts can be shown to pass into metasediments, some of high grade.

In the classic area round the N. of Lake Mjosen a non-metamorphic succession, some 1500 m thick, can be studied. The lower part, the base of which is not seen, consists dominantly of grit and conglomerate but includes, near the top, the dark Biri Limestone. Above this comes the Moelv Conglomerate, a glacogene with large, irregularly scattered boulders interbedded with the underlying and overlying sediments. The latter are thin shales, followed by the thick Quartz Sandstone. The Moelv (or Varangian) glacogene and the Quartz Conglomerate are persistent formations which can be recognised, in both unaltered and deformed states, northwards through Sweden and on to the Varanger Fjord.

The Sparagmitian is followed without any structural or metamorphic break by the Cambrian, although a disconformity is claimed in some districts. The Cambrian consists usually of a relatively thin and dominantly pelitic succession. Around Lake Mjosen the succession includes all three divisions of the Cambrian, contrasting with that of the foreland further S. around Oslo, where the Lower is missing (2.2). In the Valdres district further N. a thicker Cambrian succession of sandstones and shales occurs.

In many parts of the Scandinavian Caledonides it is possible, owing to metamorphism, to recognise only a general Cambro-Silurian succession. Fossils are, however, sufficiently plentiful in the greenschist facies to distinguish the Lower Palaeozoic subdivisions, notably in the cross-section from the Storsjon to Trondheim (see also 3.7). In the latter district the Arenig contains thick basic pillow-lavas and pyroclastics and is followed by a dominantly clastic Upper Ordovician and Lower Silurian succession, which also includes rhyolite lava horizons and a limestone at the top of the Ashgill. In the N. part of the Trondheim region Ordovician carbonate rocks become more important, and along the coast of Nordland to Bodø there are thick metadolomites and metalimestones. Sedimentary iron-ores occur at many horizons. The Llandovery, where it can be distinguished, often starts with a conglomerate, and only the Lower Silurian is present.

The Scandinavian Caledonides are intruded by numerous plutonic masses, including soda-granites or trondhjemites, gabbros, peridotites and dunites. Some of the trondhjemites show gneissose structure and some can be proved to be Lower Ordovician, indicating a synorogenic origin. Others, however, are Silurian and are, therefore, late- or post-orogenic.

Post-tectonic sandstones and conglomerates of molasse-type on Hitra Island, W. of Trondheim, range from the Downtonian to the Middle Old Red Sandstone. Further S., similar rocks in the Nordfjord-Sognefjord district have yielded only Middle Devonian fossils.

On Andøy, off the N.W. coast, a succession of sandstones and shales, with Oxfordian coal-seams near the base, ranges up to the Lower Cretaceous.

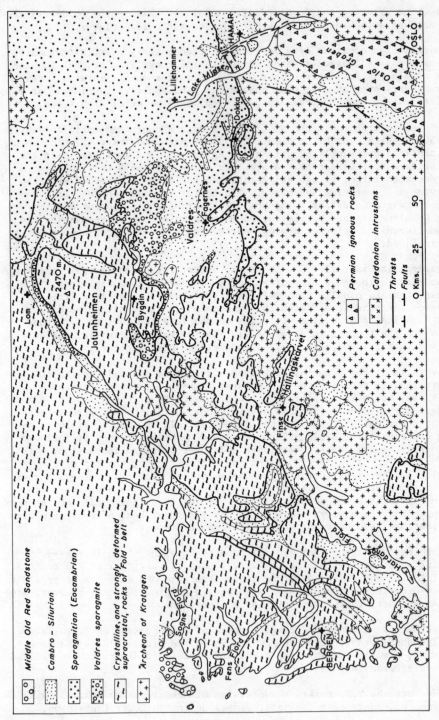

Fig. 3.15 STRUCTURAL MAP OF SOUTH CENTRAL NORWAY

The main structural events were:-

Precambrian	-	folding and metamorphism
Caledonian - Pre-Devonian	-	folding and metamorphism
Late-Caledonian - Post Middle Devonian (Svalbardian)	-	folding
Tertiary	-	faulting

Much of the fold-belt is made up of large, far-travelled nappes, themselves folded. In some districts, however, the folds are probably autochthonous. Moreover, windows revealing Precambrian "basement" are much more extensive than in the British Caledonides. The structures of four important regions will be discussed: the Jotunheim, the Trondheim-Swedish Border area, the coastal zone from Bergen to the Lofoten Islands, and Finnmark.

South-east of the Jotunheim the sparagmites form several parautochthonous nappes; W. of Lake Mjosen the Vemdal Nappe of Quartz Sandstone overrides the Cambrian of the foreland by at least 25 km. The classic sparagmite succession lies in sharp W. by N. folds formed by the sediments being pushed against the S. margin of their basin of deposition (Strand and Kulling, 1972, 26). Much of south central Norway consists largely of intermediate and basic gneisses of charnockite type with dunites. To the S. the nappe overrides the Valdres Sparagmite, containing large, drawn-out boulders, and considered to have been formed synorogenetically between two stages of nappe formation. The basal thrust-zone is displayed on the N.E. face of the Bitihorn, near Bygdin. The nappe structures were modified by N.W. cross-folding.

East of Trondheim the mainly parautochthonous Eastern Complex of the Swedish border mountains is overridden by the bottom unit of the Western Complex, the low-grade Sarv Nappe (3.7). This is overlain by the Seve/Koli Nappe Complex, consisting of high-grade, probably Precambrian, rocks overlain by low-grade Lower Palaeozoic. A number of N.-S. antiforms expose the nappe sequence progressively further E. into the Trondheim area (Trøndelag) where a major synform brings down the Trondheim Nappe of the Western Complex. Above high-grade rocks this unit also consists of greenschist facies Lower Palaeozoic, including thick Arenig volcanics (see above).

Further W. the Tomeras Antiform and a coastal belt of tight upright anti-forms and synforms shows interfolding of amphibolites, overlying dyke-intruded meta-arkoses, correlated with the Seve and Sarv Nappes respectively, with rocks which can be correlated with those of the high-grade Offerdal Nappe (3.7).

Around Bergen the well-known Arcs lie at the S. end of a major, broadly N.N.W. orientated, synform with a curved axis convex E. Tectonic interleaving before the formation of the synform has led to alternation in the Arcs of high-grade metamorphics and low-grade Cambro-Silurian fossiliferous metasediments (Strand, 1960, 227).

Further N., culminations within the Caledonides bring up large outcrops of crystalline rocks showing Precambrian structures modified by Caledonian events but, nevertheless, yielding some radiometric dates well back in the Proterozoic.

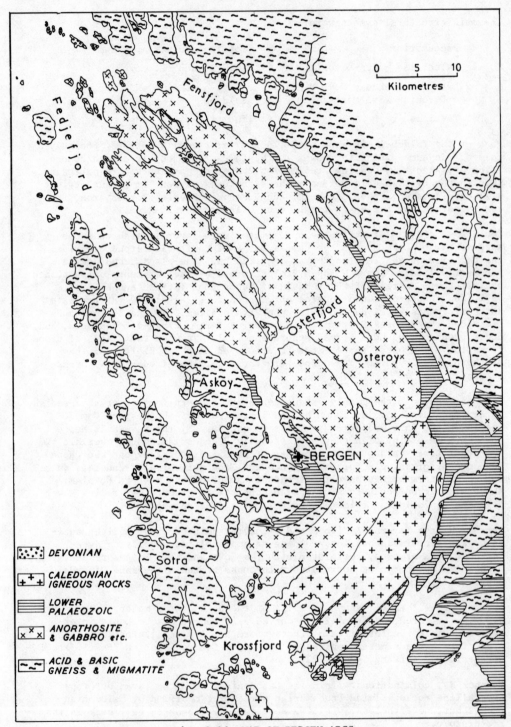

Fig. 3.16 MAP OF BERGEN ARCS

In the Lofoten Islands, and in those of the Vesterålen Group, Precambrian micaceous gneisses and amphibolites with limestones and metasedimentary iron-ores are intruded by charnockites, gabbros, anorthosites, syenites and granites, forming spectacular mountain scenery. These are probably Precambrian but some have probably been later mobilized or "Caledonised"; this would account for the inclusion of metasediments of Caledonian type. In the Tysfjord district a basal massif of granitic gneisses with mica-schist and quartzite overlies granite. The massif is brought up by W.N.W. antiforms which interfere with probably Precambrian folds on N. to N.E. axes. There is controversy regarding the age of the Tysfjord Granite. It is, however, probably Precambrian, as very similar granite to the E., both in the autochthon and in a nappe, is overlain by Cambrian.

In Finnmark the parautochthonous Gaissa Nappe (3.7) is overlain by two distinct allochthonous metamorphic nappe-complexes, the Laksefjord Nappe and the Kalak (Kolvik) Nappe. The lower Nappe has a wide outcrop round the head of Laksefjord. It consists of a greenschist facies metamorphosed sedimentary sequence, totalling at least 6 km in thickness, which can be subdivided into three formations. The lowest, Ifjord Formation, is a thick polymict conglomerate for which a glacial origin has been suggested. This is overlain by quartz sandstones of the Landersfjord Formation which show abundant sedimentary structures indicating a fluvial environment. The highest, Friarfjord Formation, consists of dark grey phyllites and interbedded siltstones which contain sedimentary structures indicative of a marine environment. The age of these sediments is not known but the complete absence of both body fossils and trace fossils strongly suggests a Precambrian age.

Five phases of deformation have been recognised in the nappe. The first phase produced asymmetric N.W. to N.-S. folds. Phyllitic cleavage was developed in pelites parallel to the axial surfaces. Pebbles in the Ifjord conglomerate were elongated in a N.N.W. direction, sub-parallel to the associated fold axes. Since the pebble deformation is much greater close to the basal thrust it is thought that significant nappe translation was associated with this first deformation phase.

The second phase produced asymmetric folds, refolding the earlier folds and phyllitic foliation about similarly trending N.-S. axes. The third deformation is characterised by small scale cross-folds of the earlier structures. Kink bands and conjugate shears were formed in the fourth phase and their association with the principal thrust surfaces suggests that final thrusting adjustments took place during this period. The last effects of the Caledonian deformation of the nappe produced broad warping, faulting and joints.

At the base of the Laksefjord Nappe a tectonic slice of basement granitoid rocks separates the main Laksefjord Nappe from the underlying Gaissa Nappe. It is thought that this represents a fragment of a basement ridge - the Finnmark Ridge (Gayer and Roberts, 1973), which separated the areas of deposition of the Dividal and Porsangerfjord Groups from the Laksefjord Group - caught up in the sole of the overriding Laksefjord Nappe.

The Kalak (Kolvik) Nappe or upper nappe has an extensive outcrop on either side of Porsangerfjord and Laksefjord. It consists of a sequence, at least 4 km thick, of upper greenschist to lower amphibolite facies metasediments, dominated by thick feldspathic psammite units interbanded with schistose pelitic and semipelitic units. Cross-bedding, ripple marks and desiccation cracks suggest a shallow marine environment (Williams, 1974). The general lithology, the sedimentary structures and the presence of calc-silicate lenses in some psammite members is reminiscent of the Moines of the Scottish Highlands (3.3). Recent investigations in the outcrop to the W. of Porsangerfjord have demonstrated the presence of three separate thrust units within the Kalak Complex (Williams et al., 1976).

On Sørøy, an island off the W. Finnmark coast, the highest thrust unit within the Kalak Complex contains a marble unit with Lower Middle Cambrian Archaeocyathids preserved. The Kalak Group, therefore, ranges from Middle Cambrian to Eocambrian in age, although the absence of Varangian Tillite within the sequence suggests that the oldest Eocambrian is not represented. The metamorphism has been dated radiometrically as 530 ± 35 m.y. (Pringle and Sturt, 1969). As in the Laksefjord Nappe, five phases of deformation have been recognised.

Windows of Precambrian "basement", with a thin veneer of tillite and Eocambrian unmetamorphosed sediments, occur to the W. and S.W. at Komagfjord and Alta. These suggest a minimum distance of postmetamorphic translation of 70 km.

Both the Laksefjord and Kalak Nappe Complexes are intruded by tholeiitic dolerite dykes which were affected by the earliest deformations. In the islands off the W. Finnmark coast a major basic and ultra-basic igneous province (the Seiland Igneous Province) was developed with intrusion throughout the deformation cycle.

Much of the island of Magerøy, to the N.W. of Porsangerfjord, is occupied by a sequence of meta-conglomerates, sandstones, shales and limestones, that contain a Llandovery fauna. This sequence is thrust over rocks of Kalak Nappe aspect and must represent an entirely younger episode of the Caledonian Orogeny, metamorphosed and deformed in a post-Llandovery event.

The northern half of Varangerhalvøya in East Finnmark is occupied by a 9.5 km thick sequence of clastic sediments - the Barents Sea Group (Siedlecka and Siedlecki, 1967) - overthrust by a second thick clastic sequence - the Raggo Group. These are separated from the autochthonous Caledonian sequence of Varangerfjord by a N.W.-S.E. trending dextral strike slip fault (Roberts, 1972). The Barents Sea and Raggo Groups are affected by only one major fold-forming deformation together with associated thrusting.

As regards the ages of folding and metamorphism of the Scandinavian Caledonides, breaks in the Ordovician succession and at the top of the Ordovician indicate a Taconic phase, supported by a number of radiometric dates around 450 m.y. There are also a few determinations of 550 m.y. (Strand and Kulling, 1972, 8). Caledonian magmatism was at its strongest in Ordovician times. A major phase was, however, post-Lower Silurian and pre-Downtonian, as Silurian sediments are metamorphosed, folded and involved in nappe-structure which does not affect the Downtonian. There is also a peak of radiometric dates at 400 m.y.

Late Caledonian (Svalbardian) movements, without metamorphism, have, however, affected the Old Red Sandstone of the W. coast. Dips of up to $50°$ occur, and faulting, both normal and reversed. Fold-axes vary from N.E. to E.

Post-Lower Cretaceous, probably Tertiary, faults cut the Mesozoic of Andøy.

Holtedahl (1960, 352) interprets the Norwegian Channel off the W. coast as a narrow, curved, Tertiary graben modified by glacial erosion.

REFERENCES

Allan, D.A. 1928. The geology of the Highland Border from Tayside to Noranside. T.R.S.E., 56, 57.
Allan, T.A. 1940. The geology of the Highland Border from Glen Almond to Glen Artney. T.R.S.E., 60, 171.
Allison, A. 1940. Loch Awe succession and tectonics: Kilmartin - Tayvallich - Danna. Q.J.G.S., 96, 423-449.
Anderson, J.G.C. 1948. The stratigraphical nomenclature of Scottish metamorphic rocks. Geol. Mag., 85, 89-96.
Anderson, J.G.C. 1956. The Moinian and Dalradian rocks between Glen Roy and the Monadliath Mountains, Inverness-shire. T.R.S.E., 63, 15-36.
Anderson, J.G.C. 1960. The Wenlock Strata of South Mayo. Geol. Mag., 97, 265-275.
Anderson, J.G.C. 1965. The Precambrian of the British Isles. The Precambrian, vol. 2, (ed. K. Rankama), 25-111. New York.
Asklund, B. 1960. Studies in the Thrust Region of the Southern Part of the Swedish Mountain Chain. Guide to excursions A24 and C19, I.G.C. 21 Sess. Norden.
Badley, M.E. 1976. Stratigraphy, structure and metamorphism of Dalradian rocks of the Maumturk Mountains, Connemara, Ireland. Journ. Geol. Soc., 132, 509-520.
Bailey, E.B. 1922. The Structure of the South-West Highlands of Scotland. Q.J.G.S., 68, 82-131.
Bailey, E.B. & Weir, J. 1933. Submarine faulting in Kimmeridgian times, East Sutherland. T.R.S.E., 57, 429-467.
Barber, A.J. & May, F. 1976. The history of the Western Lewisian in the Glenelg Inlier, Lochalsh, Northern Highlands. Scot. Journ. Geol., 12, 35-50.
Barber, A.J. & Soper, N.J. 1973. Summer Field Meeting in the North-West of Scotland, 1971. P.G.A., 84, 207-235.
Barrow, G. 1893. On the intrusion of muscovite-biotite gneiss in the south-eastern Highlands of Scotland, and its accompanying metamorphism. Q.J.G.S., 49, 330.
Bassett, D.A. 1958. Geological Excursion Guide to the Glasgow District. Geol. Soc. Glasgow.
Bickle, M.J. et al. 1972. The Silurian of the Croagh Patrick Range, Co. Mayo. Sci. Proc. Roy. Dublin Soc., A4, 231-249.
Christie, J.M. 1963. The Moine Thrust-Zone in the Assynt Region, North-West Scotland. Univ. of California Publications in Geol. Sci., 40, 345-440.
Clifford, T.N. 1957. The stratigraphy and structure of part of the Kintail district of southern Ross-shire; its relation to the Northern Highlands. Q.J.G.S., 113, 57-92.
Clough, C.T., Maufe, H.B. & Bailey, E.B. 1909. The Cauldron subsidence of Glen Coe and the associated igneous phenomena. Q.J.G.S., 65, 611.
Dalziel, I.W.D. & Brown, R.L. 1965. The structural dating of the sillimanite grade metamorphism of the Moines in Ardgair (Argyll) and Moidart (Inverness-shire). Scot. Journ. Geol., 1, 304-311.
Dalziel, I.W.D. & Johnson, M.R.W. 1963. Evidence for the geological dating of the granitic gneiss of Western Ardgair. Geol. Mag., 100, 244-254.
Downie, C. et al. 1971. A palynological investigation of the Dalradian rocks of Scotland. I.G.S. Rep. 71/9.

Flinn, D. 1967. The metamorphic rocks of the southern part of the Mainland of Shetland. Geol. Jnl., 5, 251-290.

Flinn, D. 1977. Shetland faults. Journ. Geol. Soc., 133, 231-248.

Føyn, S. 1967. Dividal-Gruppen ("Hyolithus-sonen") i Finnmark og dens forhold til de en kambrish-kambriske formasjones. Norges Geologiske Undersøkelse, 249, 1-84.

Gayer, R.A. & Roberts, J.D. 1971. The structural relationships of the Caledonian nappe of Porsangerfjord, West Finnmark, N. Norway. Norges Geologiske Undersøkelse, 269, 21-67.

Gayer, R.A. & Roberts, J.D. 1973. Stratigraphic Review of the Finnmark Caledonides with Possible Tectonic Implications. P.G.A., 84, 405-428.

Gee, D.G. 1975. A Geotraverse through the Scandinavian Caledonides - Ostersund to Trondheim. Sveriges Geologiska Undersoking CNR717.

Gee, D.G. (ed.). 1976. Geotraverse Excursion (Ostersund to Trondheim). Swedish Geodynamics Project. Caledonian Research Project. (rev. ed.).

Harris, A.L. 1963. Structural investigations in the Dalradian rocks between Pitlochry and Blair Atholl. Trans. Edin. Geol. Soc., 19, 156-178.

Harris, A.L., Bradbury, H.J. & McGonigal, M.H. 1976. The evolution and transport of the Tay Nappe. Scot. J. Geol., 12, 103-113.

Harris, A.L. & Pitcher, W.S. 1975. The Dalradian Supergroup. Geol. Soc. Spec. Rep. No. 6, 52-75.

Harrison, V.E. & Moorhouse, S.J. 1976. A possible early Scourian supracrustal assemblage within the Moine. Journ. Geol. Soc., 132, 461-466.

Holtedahl, O. (ed.). 1960. Geology of Norway. Norges Geologiske Undersøkelse, NR208.

Holtedahl, O. et al. 1960. Aspects of the Geology of Northern Norway. Internat. Geol. Congr. Norden 1960, Exc. Guide A3.

Johnson, M.R.W. 1960. The structural history of the Moine Thrust-Zone at Lochcarron, West Ross. T.R.S.E., 64, 139.

Johnson, M.R.W. 1969. Dalradian of Scotland. In: North Atlantic - Geology and Continental Drift. (ed. M. Kay). Am. Soc. Pet. Geol.

Johnstone, G.S. et al. 1969. Moinian Assemblage of Scotland. North Atlantic - geology and continental drift, 159-180. Amer. Assoc. Petrol. Geologists.

Kennedy, W.Q. 1946. The Great Glen Fault. Q.J.G.S., 102, 41-76.

Kvale, A. 1960. The Nappe Area of the Caledonides in Western Norway. Internat. Geol. Congr. Norden 1960. Exc. Guide A7 and C4.

Lambert, R.St.J. & Poole, A.B. 1964. Guide to the Moine Schists and Lewisian Gneisses around Mallaig, Inverness-shire. Geol. Assoc. Guide No. 35.

Lapworth, C. 1883. The secret of the Highlands. G.M., 10, 120, 193, 337.

MacGregor, A.G. 1967. Faults and Fractures in Ardnamurchan, Moidart, Sunait, and Morvern. Bull. Geol. Surv. G.B., 27, 1-15.

Macgregor, M. & Phemister, J. 1972. Geological Excursion Guide to the Assynt District of Sutherland. (3rd ed.). Edin. Geol. Soc.

May, F. 1970. Movement, Metamorphism and Migmatization in the Scallaway Region of Shetland. Bull. Geol. Surv. G.B., 31, 205-226.

McKerrow, W.S. & Campbell, C.J. 1960. The Stratigraphy and Structure of the Lower Palaeozoic Rocks of North-West Galway. Sci. Proc. Roy. Dublin Soc., Ser. A., 1, 27-52.

Mykura, W. 1976. British Regional Geology: Orkney and Shetland. N.E.R.C. I.G.S. H.M. Stat. Office.

Peach, B.N. & Horne, J. 1884. Report on the Geology of the North-West of Sutherland. Nat., 31, 31.

Peach, B.N. et al. 1907. The Geological Structure of the North-West Highlands of Scotland. Mem. Geol. Surv.

Peach, C.W. 1854. Notice of the discovery of fossils in the limestones of Durness in the County of Sutherland. Edinburgh New Phil. Journ. (N.S.), 2, 197.
Phillips, W.E.A. 1973. The pre-Silurian rocks of Clare Island, Co. Mayo, Ireland, and the age of the metamorphism of the Dalradian in Ireland. Journ. Geol. Soc., 129, 585-606.
Phillips, W.E.A. 1974. The stratigraphy, sedimentary environments and palaeogeography of the Silurian strata of Clare Island, Co. Mayo, Ireland. Journ. Geol. Soc., 130, 19-41.
Piasecki, M.A.J. 1975. Tectonic and metamorphic history of the Upper Findhorn, Inverness-shire, Scotland. Scot. Journ. Geol., 11, 87-115.
Pidgeon, R.T. 1969. Zircon U-Pb ages from the Galway Granite and the Dalradian, Connemara, Ireland. Scot. Journ. Geol., 5, 375-392.
Pitcher, W.S. & Cheesman, R.L. 1954. Summer Field Meeting in North-West Ireland, with an introductory note on the geology. P.G.A., 65, 345-371.
Pitcher, W.S. & Read, H.H. 1960. The aureole of the main Donegal Granite. Q.J.G.S., 116, 1-34.
Pitcher, W.S. et al. 1964. The Leannan fault. Q.J.G.S., 120, 241-273.
Pitcher, W.S. et al. 1971. The Ballybofey Anticline: a solution of the general structure of parts of Donegal and Tyrone. Geol. Journ., 7, 321-328.
Poole, A.B. & Spring, J.S. 1974. Major structures in Morar and Knoydart, N.W. Scotland. Journ. Geol. Soc., 130, 43-53.
Powell, D. 1974. Stratigraphy and structure of the Western Moine and the problem of Moine orogenesis. Journ. Geol. Soc., 130, 575-593.
Pringle, J.R. 1970. The structural geology of the North Roe area of Shetland. Geol. Journ., 7, 147-170.
Pringle, J.R. & Sturt, B.A. 1969. The Age of the Peak of the Caledonian Orogeny in West Finnmark, North Norway. Norges Geologiska Undersøkelse, 49, 435.
Ramsay, J.G. 1958. Superimposed folding at Loch Morar, Inverness-shire and Ross-shire. Q.J.G.S., 113, 271-305.
Read, H.H. 1960. North-East Scotland. The Dalradian. Geol. Assoc. Exc. Guide No. 31.
Richey, J.E. & Kennedy, W.Q. 1939. The Moine and sub-Moine Series of Morar, Western Inverness-shire. Bull. Geol. Surv., 2, 26.
Riddihough, R.P. & Max, M.D. 1976. A geological framework for the continental margin to the west of Ireland. Geol. Journ., 2, 109-120.
Roberts, D. 1972. Tectonic Deformation in the Barents Sea Region of the Varanger Peninsula, Finnmark. Norges Geologiska Undersøkelse, 282, 1-39.
Roberts, J.L. 1974. The Structure of the Dalradian Rocks in the South-West Highlands of Scotland. Journ. Geol. Soc., 130, 93-124.
Rosendahl, H. 1945. Prekambrian-Eokambrian i Finnmark. Norges Geologiske Undersøkelse, 25, 327-349.
Shackleton, R.M. 1958. Downward-facing structures of the Highland Border. Q.J.G.S., 113, 361-392.
Siedlecka, A. & Siedlecki, S. 1967. Some new aspects of the geology of Varanger Peninsula (Northern Norway). Norges Geologiska Undersøkelse, 247, 288-306.
Soper, N.J. & Wilkinson, P. 1975. The Moine Thrust and Moine Nappe at Loch Eriboll, Sutherland. Scot. Journ. Geol., 11, 339-359.
Strand, T. 1960. In: Geology of Norway. (ed. O. Holtedahl). Norges Geologiske Undersøkelse, NR208.
Strand, T. & Holmsen, P. Stratigraphy, Petrology and Caledonian Nappe Tectonics of Central Southern Norway; Caledonoid Basal Gneisses in a North-Western Area (Offdal-Sunndal). Internat. Geol. Congr. Norden 1960,

Exc. Guide A13 and C9.

Strand, T. & Kulling, O. 1972. Scandinavian Caledonides. Wiley, London and New York.

Stromberg, A. 1955. Zum gebirgsbau der skanden i mattleren Harjedalen. Bull. geol. Inst. Uppsala, 35, 199-243.

Sutton, J.S. 1972. The Pre-Caledonian Rocks of the Mullet Peninsula, County Mayo, Ireland. Roy. Dublin Soc., 4A, 121-136.

Sutton, J. & Watson, J. 1954. The structure and stratigraphical succession of the Moines of Fannich Forest and Strathbran. Q.J.G.S., 110, 21-53.

Sutton, J. & Watson, J. 1958. Structures in the Caledonides between Loch Duich and Glenelg, North-West Highlands. Q.J.G.S., 114, 231-257.

Tanner, P.W.G. 1970. The Sgurr Beag Slide - a major tectonic break within the Moinian of the Western Highlands of Scotland. Q.J.G.S., 126, 435-463.

Thorslund, P. & Jaanusson, V. 1960. The Cambrian, Ordovician and Silurian in Vastergorland, Narke, Dalarna and Jamtland, central Sweden. Guide to excursions A33 and C18, IGC, 21 Sess. Norden.

Van Breeman, O. et al. 1974. Precambrian and Palaeozoic pegmatites in the northern Moines of Scotland. Journ. Geol. Soc., 130, 493-508.

Voll, G. 1964. Deckenbau und fazies im Schottischen Dalradian. Geol. Rundschau, 53, 590-612.

Watson, J. 1975. The Lewisian Complex. Geol. Soc. Spec. Rep. No. 6, 15-29.

Westoll, T.S. 1964. The Old Red Sandstone of north-eastern Scotland. Adv. Sci., 20, 446.

Williams, G.D. 1974. Sedimentary Structures in the amphibolite facies rocks of the Bekkarfjord formation, Laksefjord, Finnmark. Norges geologiska Undersøkelse, 311, 35-48.

Williams, G.D. et al. 1976. A revised tectono-stratigraphy of the Kalak Nappe in Central Finnmark. Norges geologiska Undersøkelse, 324, 47-61.

Yardley, B.W.D. 1976. Deformation and metamorphism of Dalradian rocks and the evolution of the Connemara cordillera. Journ. Geol. Soc., 132, 521-542.

Chapter 4
CALEDONIAN FOLD-BELTS (MAINLY SEDIMENTARY)

4.1 Introduction

South of the Highland Boundary Fracture-Zone much of the British Isles belongs to a part of the Caledonian Fold-Belt in which Caledonian regional metamorphism did not go beyond the development of cleavage (apart from Connemara and South Mayo in Western Ireland (3.6)). Large sections of this fold-belt are hidden under a younger cover. Major Caledonian blocks come to the surface in Wales and the Welsh Borderland (4.2), the Lake District and the Isle of Man (4.3), the Southern Uplands of Scotland and their continuation into Ireland (4.4) and South-East Ireland (4.5). In all these blocks there are post-tectonic Caledonian plutons which have thermally metamorphosed the adjacent Lower Palaeozoic. All the blocks, too, have been modified by Hercynian and Tertiary events.

4.2 Wales and the Welsh Borderland (Figs. 4.1, 4.2 and 4.3)

This is the type region for the Lower Palaeozoic System, Cambria being the Latin name for Wales and the Ordovices and Silures tribes of the Welsh Borderland in Roman times. In North-West Wales, where the Ordovician contains resistant volcanics, picturesque, rugged mountains rise to over 3000 ft (915 m), culminating in Snowdon 3560 ft (1085 m). In Central Wales, where Silurian sediments predominate, rounded hills with fewer outcrops rise to over 2000 ft (610 m). Nearly all the valleys of the region are strongly glaciated, and thick glacial deposits cloak much of the lower ground. Excellent exposures occur along the W. and N. coasts and around the large, low island of Anglesey. Much of the eastern margin and the southern margin are determined by the Church Stretton - Careg Cennen line (see below) near which there are also good exposures, especially in the Church Stretton area.

Excursions in parts of the region are described by Dean (1968) and Williams and Ramsay (1959).

The formations present are:-

> Tertiary
> Jurassic
> Triassic
> Carboniferous
> Lower Old Red Sandstone
> Silurian
> Ordovician
> Cambrian
> Precambrian

The region contains the most extensive development of the Precambrian in the British Isles outside Scotland and Ireland. In the E., near Church Stretton, the following succession has been established:-

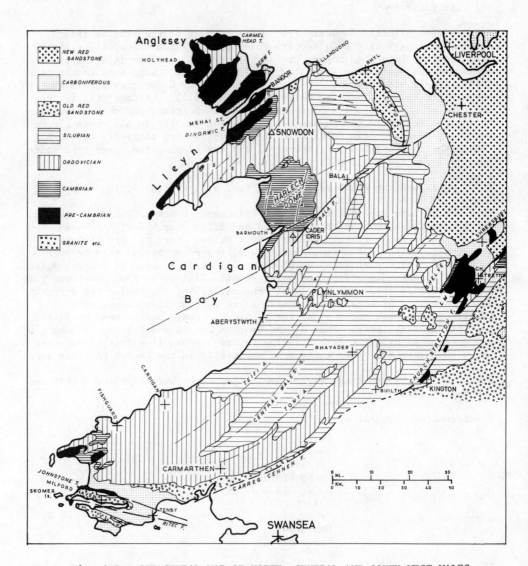

Fig. 4.1. STRUCTURAL MAP OF NORTH, CENTRAL AND SOUTH-WEST WALES

Western Longmyndian or Wentnorian
    (unconformity)
Eastern Longmyndian or Strettonian
    (unconformity)
Uriconian.

The Uriconian, which also outcrops in Central England (6.4), consists of rhyolite lava and pyroclastics with some basalts. The unconformably overlying Stretton Group contains shales (often cleaved), flagstones and grits, with thin volcanics. It is followed by the Wentnor Group of red or purple arkoses, conglomerates and shales which have been compared with those of the Torridonian (2.3).

In the S.W. of the region, in Pembrokeshire, the Pebidean volcanics and intrusions, separated by a major erosional phase from the Cambrian, can be correlated with the Uriconian. Acid volcanics and pyroclastics between Snowdon and Bangor, on the other hand, pass conformably upwards by intercalation into basal Cambrian conglomerates and grits and are, therefore, probably younger than the Uriconian. Further W., the Monian forms much of Anglesey and the Lleyn Peninsula to the S. This formation consists of over 20,000 ft (6100 m) of quartzites, schistose greywackes, muscovite-chlorite-schists, spilites with pillow structure and pyroclastics. The Mona Complex is considered to be late Precambrian and to have undergone late Precambrian metamorphism (Moorbath & Shackleton, 1966). The oldest overlying rocks are Arenig.

A curious feature of the Welsh Block is that no glacogene has been discovered in the late Precambrian, such as those of the Dalradian (3.2) of Scotland and Ireland and the Brioverian of Brittany (5.9).

The thickest (15,000 ft; 4600 m) and most complete Cambrian succession occurs in the Harlech Dome. It consists of grits or greywackes (turbidites with large-scale graded bedding) and shales; the base is not seen and the first fossils occur well above the lowest beds, which may be Precambrian. In the N. and in the S.W. of the region much thinner, less complete Cambrian successions are present. In the N. Cambrian slates have been worked on a vast scale in a belt extending N.N.E. from Dinorwic, near Snowdon. Intermittent movements of a marginal shelf area during the Cambrian are exemplified by breaks in the succession S. of Shrewsbury, especially at the base and top of the Middle Cambrian.

The Ordovician was a period of intense volcanic activity, which included a number of ignimbrites. The Arenig rests unconformably on the Monian in Anglesey and in other districts oversteps onto various parts of the Cambrian succession. Arenig deposits in the Welsh Geosyncline begin with shallow-water conglomerates, grits and quartzites but gradually give way upwards to flags or shales in the higher Arenig and in the Llanvirn. Vulcanicity, mostly submarine, was intense in the Cader Idris, Arenig, Manod and Moelwyn areas of North Wales, and in the Skomer, Fishguard, Trefgarn and Builth areas further S. In Shropshire, vulcanicity was more limited, occurring in the Llanvirn of Shelve. Llandeilian deposits are probably confined in North Wales to the Berwyn Dome (where they comprise 2000 ft of shelly facies silts, sands and calcareous sediments, interspersed with acid tuffs and flows). In South Wales, and again in Shelve, the Llandeilian consists predominantly of calcareous flags, vulcanicity being restricted.

Mid-Ordovician uplift and erosion was thus relatively extensive and the (localized) intensity of these movements is shown by the rapid Bala overstep on the W. margin of the Longmynd and again near Tremadoc where a local anticline, over 2000 ft high, was removed before Caradocian deposition began. Bala deposition was relatively quiet in the S. of the region, but was interrupted by violent vulcanicity in Caernarvonshire. The eruptions, not always submarine, extruded volcanics ranging from basic to acid (rhyolites predominating) but probably all derivatives of a basaltic magma. Dipping E. from the Dinorwic Cambrian slate-belt, the Lower Ordovician consists of thick graded grits or subgreywackes and slates, followed by the Snowdon syncline of Caradoc acid lavas and pyroclastics. Cleavage has developed workable slates in the Ordovician pelites.

In general, the Bala sediments of North Wales were of basin facies in the W. and N. but of more marginal or shelf type, with intercalated limestones, further E. In South Wales the Towy Anticline appears to have frequently separated the two major Ordovician facies, and in Ashgillian times formed an effective barrier with no deposition to the E. Breaks in Bala sedimentation occur in the Bala district, the West Berwyns and along the Towy Anticline, whilst in the Beddgelert area, mid-Caradocian faulting influenced the position of volcanic piles (Rast, 1961). In the N.W. part of Snowdonia Beavon (1963) suggested that a local Lower Caradoc intra-volcanic unconformity is part of a buried volcanic collapse-structure.

At the beginning of the Silurian, yet another phase of unrest occurred, these movements reaching considerable intensity on the eastern border of the Welsh Geosyncline but probably dying away inwards. As a result the E. edge of the Welsh Geosyncline, which had been over the Longmynd in Caradoc times, moved westwards to the Builth area. The Silurian consists mainly of shales or mudstones, silts, sandstones or subgreywackes (often graded turbidites), grits and conglomerates; cleavage has developed in the finer beds producing locally workable slates.

Old Red Sandstone "molasse" forms much of the S.E. and S. margin of the region but is sparsely preserved within the fold-belt. Some 4000 ft (1220 m) of the formation rests on Silurian in the Clun area in the E., and in Anglesey there are 1300 ft (395 m) of Lower Old Red conglomerates, sandstones and marls. The Carboniferous is also restricted; the succession is similar to that in nearby parts of Central England (6.4).

The Triassic, confined to the N., consists of a considerable thickness of sandstones and marls. The Mochras bore, on the coast N. of Aberystwyth, proved a thick L. Jurassic succession under Tertiary W. of a major N.-S. fault downthrowing towards Cardigan Bay.

Structural Evolution

The main structural events were :-
                                      Precambrian

                                      (Post-Cambrian/Pre-Arenig
                                      (Mid-Ordovician
            Caledonian   (Post-Ordovician/Pre-Silurian
                                      (Late Silurian
                                      (Mid-Devonian

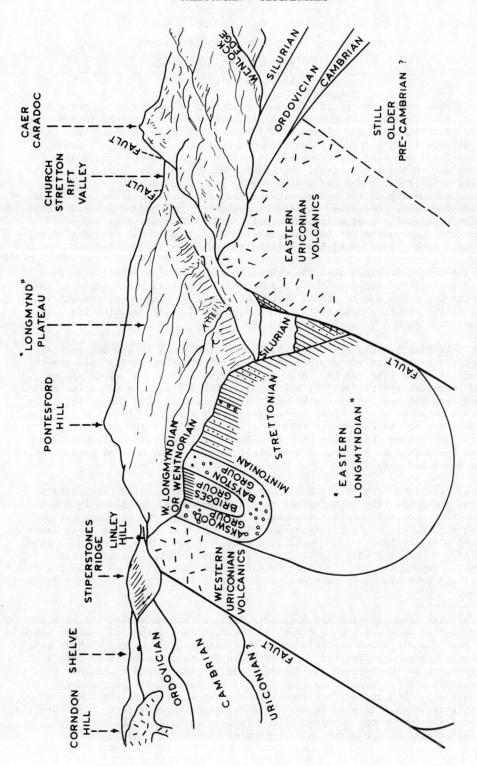

Fig. 4.2. THE STRUCTURE OF THE LONGMYND, SHROPSHIRE (SALOP)

Hercynian   Post-Westphalian (Asturian)
Alpine (sensu lato)
Pliocene - uplift

The Monian metasediments of Anglesey, placed in stratigraphical order by Shackleton (1954), form complex N.E. striking folds overturned towards the S.E. Both flexural and cleavage folding were involved in this first phase of deformation. A second phase produced conjugate sets of small-scale shear-folds and also cross-folds. All the structures were linked by a lineation in the a direction. Metamorphism was operative during the main (first) deformation and was waning during the second phase. Locally the rocks became migmatized and granitized, and then were injected by a potash granitic magma. One other structural feature of the Bedded Series is the Mélange extending over many square miles and consisting of masses of quartzite, greywacke, pillow-lava, mostly belonging to the youngest (Gwna) group. Such a mass could have been bulldozed under a giant nappe.

In the Church Stretton area the Uriconian and overlying Strettonian were folded and then eroded to the W., so that on this side the Wentnorian rests on the Uriconian and to the E. on the Strettonian. The whole Precambrian succession was then compressed into a deep N.N.E. syncline with its axial plane dipping W.N.W. at $70°$ (James, 1956).

The fold lies immediately W. of the Church Stretton line and may indicate that the latter was already a zone of crustal weakness in the late Precambrian. The line can be traced from the Wrekin S.W. to beyond Kington, and there are several smaller exposures of the Precambrian along its course including Wentnorian-type rocks near Knighton and igneous rocks, which may be Uriconian or may be older Precambrian, near Kington. The line marks the S.E. front of the Caledonian fold-belt, a conclusion based not only on structure but on separation of the thick Welsh Lower Palaeozoic succession from the much less folded, thinner shelf succession of Central England (6.4) and South Wales (6.5).

Faults inclined towards the N.W., which were probably Caledonian thrusts, occur along the line, but further Hercynian movement, sometimes in the reverse sense, and possibly steepening has occurred. In the Caer Caradoc district, E. of Church Stretton, Cobbold proved a complex zone of at least three parallel faults (two vertical and one central fault with a N.W. hade of $45°$). Only the westernmost fracture now cuts Carboniferous or later rocks. Beyond Kington the Careg Cennen Disturbance, although steep and with a strong Hercynian component, probably continues the Church Stretton fault-zone.

Caledonian trends, which are N.E. to N.N.E., with some variations in North and Central Wales, swing to E.N.E. in the S.W. The climax of the movements in most districts was probably just after the Silurian as the highest beds of this System are affected. In the Lleyn Peninsula there is an angular difference of $10°$ between the Cambrian and the Arenig and variations in the amount of the pre-Arenig gap in Snowdonia suggested that the Snowdon Synclinorium and the Harlech Dome were already active. In the S.W. the Arenig rests on Middle Cambrian. Further movement occurred in Mid-Ordovician times as shown by the overstep of the Caradoc in several areas; early Silurian disturbance seems to have been marked only in the E.

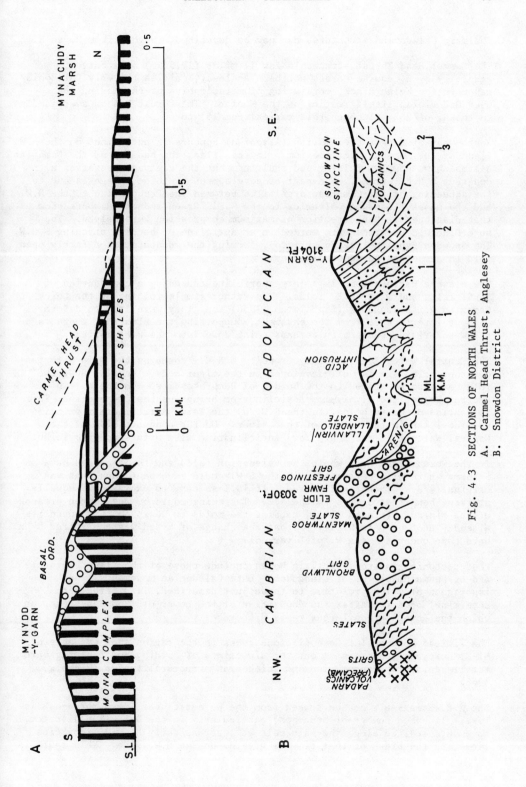

Fig. 4.3 SECTIONS OF NORTH WALES
A. Carmel Head Thrust, Anglesey
B. Snowdon District

The main Caledonian structures can now be described, roughly from N. to S.

The Carmel Head Thrust, traceable for 11 miles (17.5 km) on a curved, generally E.-W. course across northern Anglesey, carries the only Caledonian nappe in the Welsh Block, consisting of metamorphics of the Monian pushed over Ordovician itself resting on the Monian. The thrust has an average dip to the N. of $25°$, although this varies from $15°$ to $60°$.

South-west of Bangor Precambrian (Arvonian) appears in anticlines on the N.W. flank of the Snowdon Synclinorium. In this flank the bedding of the Cambrian slates shows complex folding and faulting, but the cleavage preserves a constant N.N.E. strike. The perfect development of the cleavage is due to the squeezing of the Cambrian argillites between the Precambrian to the N.W. and the massive Snowdon volcanics to the S.E. Green reduction spots show that distances in the direction of maximum compression were halved. The huge Dinorwic Quarries were worked in an anticline of bedding pitching S.S.W. The Snowdon Syncline (sensu stricto), forming the mountain, is a fairly open fold in competent Caradoc volcanics.

The classic Harlech Dome is a large pericline caused by a combination of N.-S. folds and later N.E. folds. The rather simple folding of the thick Cambrian grits suggests (Shackleton, 1958) that they were free to deform without interference from the basement, supporting the view that there is a thick underlying succession perhaps going down into the Precambrian.

The Central Wales Synclinorium, trending N.N.E., contains the main Silurian outcrop; cleavage is well-developed in the finer beds. Bending S.W. the fold can also be traced to the border of Pembrokeshire. To the W., in Central Wales, the Plynlymmon Anticlinorium brings up the Ordovician. This is continued, en echelon, to the S.W. by the Teifi Anticlinorium bringing up Ordovician in the valley of that river. On the opposite side of the Central Wales fold the sharp Towy Anticlinorium also brings up Ordovician.

In the Welsh Block as a whole the Caledonian folds and the cleavage have an S-form. Hercynian deformation in the S.W. could have bent back what was originally S.W. trending flexures. It is less easy to explain the equally striking bend that occurs in the N.E. The swing could be due to the presence of the Midland kratogen (6.4), causing the folds to begin to turn round its N. end. Another possibility is that the change of trend was controlled by a hard core under the sea W. of Liverpool.

Post-tectonic granites, etc., in Wales include those of the Lleyn Peninsula and of Penmaenmawr, W. of Llandudno. Late Caledonian movements (the importance of which, relative to those just described, is difficult to determine) occurred after the deposition of the Lower Old Red Sandstone, on which the Lower Carboniferous rests discordantly.

The limited outcrops of Carboniferous rocks in the region show evidence of N.-S. Hercynian compression but the directions of folding and faulting were totally guided by Precambrian and Caledonian structural trends (Nichols, 1968).

The N.E. Bala Fault can be traced from the W. coast near Towyn through Bala Lake. It has a number of branches, particularly in the Dolgellau district. Movements started along the main belt at least as early as mid-Ordovician times, as the break of that time is greater on the N.W. side. This fault is

the most continuous of a very large number of roughly strike faults, many of
which have post-Caledonian components. Thus the N.E. Berw Fault in S.E.
Anglesey separates Carboniferous from Monian and Ordovician. The parallel
Dinorwic Fault along the Menai Straits brings Carboniferous against Arvonian
volcanics with N.W. downthrow. This line, however, appears to have been one
of S.E. downthrow in pre-Arenig times (Shackleton, 1958). Earthquakes
occurred here in 1903 and 1906. Cross-faults are also abundant.

The N.W. striking Llanberis Fault has a dextral wrench component of one km
where it cuts the Cambrian slate-belt, but dies out rapidly S.E. In Central
Wales, N.W., W.N.W. and W. faults are mineralised, mainly with galena and
zincblende, which were extensively mined at one time. The age of the
mineralisation is probably Alpine or at any rate due to Alpine re-mobilisation.

Post-Triassic faulting formed the graben of the Vale of Clwyd and post-
Jurassic faulting occurred along the Cardigan Bay coast. Pliocene uplift
caused tilting and elevation of Neogene platforms.

4.3   The Lake District and the Isle of Man   (Figs. 4.4 and 4.5)

The Lake District proper consists mainly of Lower Palaeozoic rocks forming
the heart of a Tertiary dome. These rocks, and particularly the Borrowdale
Volcanic Series (Llanvirn/Llandeilo), have been eroded into rugged scenery
culminating in Scafell Pikes, 3210 ft (980m), the highest point in England.
The radiating valleys of the dome are highly glaciated, the lakes occupying
rock-basins. Exposures are excellent in many parts. Excursions are
described by Williamson (1967), Mitchell (1970) and Mitchell and others
(1972).

Structurally, the dome affects a wide area; a description is given in the
present section of the district between the Solway Firth and Morecambe Bay,
bounded to the E. by the Pennine Fault. The Isle of Man is also included.
The succession is as follows :-

               Liassic
               Triassic
               Permian
               Carboniferous
               ?Lower Old Red Sandstone
               Silurian
               Ordovician
               Cambrian.

The Manx Slate Series, making up most of the Isle of Man, is probably all of
Cambrian age (Simpson, 1963).

The Skiddaw Slates, a formation including grits, flagstones and highly
cleaved shales or mudstones, form the northern part of the Lake District
proper, with detached areas at Black Combe and near Ullswater. They probably
belong to the Arenig and Lower Llanvirn, although it has been claimed that
the lower part includes Cambrian strata. The main outcrop is followed to
the S. by the Borrowdale Volcanic Series of andesitic and rhyolitic-lavas,
agglomerates and tuffs. These unfossiliferous rocks are held to belong to
the Llanvirn and Llandeilo Series. Some of the rhyolites have been claimed
to be intrusions. The Caradoc is represented by the Coniston Limestone

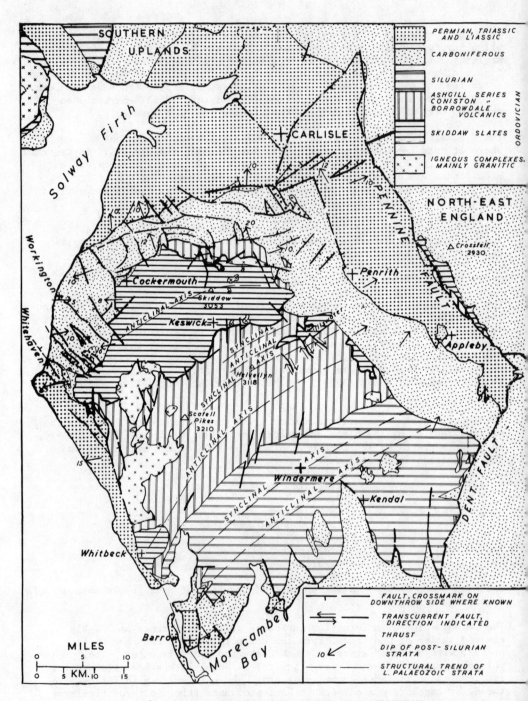

Fig. 4.4 STRUCTURAL MAP OF THE LAKE DISTRICT

Series which consists of conglomerates, limestones, calcareous shales, ashes and rhyolite-lava. The Ashgill Series starts with a thin limestone followed by shales. The Silurian consists entirely of sedimentary rocks placed in a number of lithological groups, which palaeontological evidence suggests make up nearly the full succession.

The Lower Palaeozoic rocks are cut by a number of post-tectonic, Caledonian acid to ultrabasic plutonic and hypabyssal intrusions. Around the complex Carrock Fell intrusion there is considerable and varied mineralisation which must have involved mobilisation as dates are Hercynian or even later.

Overlying the Lower Palaeozoic and beneath beds of undoubted Lower Carboniferous age there are often conglomerates and grits, the younger of which are probably basal Carboniferous, but the older probably Lower Old Red Sandstone. These are followed by thick limestones and dolomites overlain by shales and sandstones. The Millstone Grit is poorly developed. There is, however, a fairly full Coal Measure succession in the N.W. which includes the Cumberland Coalfield. The highest horizons of the Coal Measures are, however, missing because of pre-Permian denudation.

The Permian consists of red sandstones with "millet-seed" grains (the Penrith Sandstone) of breccias or "brockrams" and of discontinuous beds of Magnesian Limestone. The Triassic contains red shales and sandstones showing much lateral variation; in fact, the boundary between the Permian and Triassic is largely arbitrary. Near Carlisle dark shales and argillaceous limestone of Lower Lias age are present. Lower Carboniferous, Permian and Triassic strata also occur in the Isle of Man.

Structural Evolution

Structural events were :-

Caledonian	(Mid-Ordovician
	(Late-Silurian
	(? Mid-Devonian
Hercynian	(Pre-Namurian (Sudetic)
	(Post-Westphalian (Asturian)
	Alpine (_sensu_ _lato_).

The unconformity at the base of the Coniston Limestone shows that important movements of pre-Caradoc date took place after the deposition of the Borrowdale Volcanic Series. The main structures of the Lake District, produced by compression of probably late-Silurian age, trend from N.E. (in the S.W.) to E.N.E. In the N. the Skiddaw slates occur in a broad anticlinorium within which the Loweswater Flags (the oldest group, with no exposed base) come up in three or four anticlinal zones with younger groups occupying intervening closely packed synclines. The presence of Borrowdale Volcanics in a narrow outcrop on the N. margin of the anticlinorium may be evidence for the oncoming of a Lower Palaeozoic synclinal-zone under the unconformable Carboniferous.

The northerly (Binney-Eycott) succession of the volcanics is probably conformable on the Skiddaw Slates (Moseley, 1975). However, there has been a great deal of debate regarding the relationship of the southerly Borrowdale Volcanics to the Slates. The junction is frequently obscured by

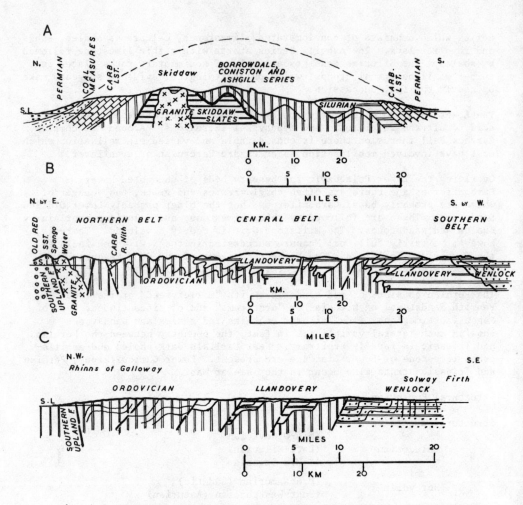

Fig. 4.5 SECTIONS ACROSS THE LAKE DISTRICT AND THE SOUTHERN UPLANDS
A. Lake District
B. Southern Uplands (after Peach and Horne)
C. Southern Uplands (after Walton)

drift but in places appears to be a thrust dipping S.E., e.g. at Ullswater. Moseley (1975) now considers that the junction is an angular but non-orogenic unconformity over the greater part of the S. outcrop.

The folding of the two groups is disharmonic, as the tight-folding of the relatively incompetent Skiddaw Slates contrasts with the open folding of the more massive volcanics. Under these circumstances movement at the junction is inevitable.

The most important structures of the Borrowdale Volcanics are a syncline passing through Scafell and Helvellyn and an anticline to the S.E. through Langdale. Further S.W. near Whitbeck, on the continuation of this upfold, there is an outcrop of Skiddaw Slates forming Black Combe. To the S. of the

Langdale Anticline the Silurian occurs in a broad synclinal zone complicated by subsidiary flexures.

The finer-grained Lower Palaeozoic rocks, particularly the shales and tuffs, show strong cleavage. Its relationship to the axes of the main folds is not always simple; it is probable that the late-Silurian folding involved two or more differing stress-phases. In the Helvellyn area cleavage is parallel to the axial planes of the folds, and S.E. of Ullswater steep cleavage also strikes parallel to the synclinal fold. In the Dunnerdale district, on the other hand, Mitchell found that the cleavage dips generally N.N.W. agreeing neither with the south-easterly dip of the Coniston Limestone or the north-easterly strike of the Borrowdale rocks.

The Ordovician and Silurian rocks are cut by a number of faults, dominant directions being N.W. and N.N.E.; near Ullswater there are also N.-S. and E.-W. fault-trends (Moseley, 1964). These faults may have been initiated during the Caledonian movements, although there are probably also Hercynian and Tertiary components.

In the Isle of Man, Simpson (1963) has shown that the earliest structure is a major north-easterly trending syncline with vertical axial plane; this was followed by a second phase of more open folds, the axial planes of which have the same Caledonoid strike but dip gently N.W.; thirdly, there were large open cross-flexures with axial planes dipping steeply E.N.E. After the last fold-phase, a renewal of compression caused sinistral transcurrent movement along N.W. or N.N.W. faults.

The absence of proved Upper Old Red Sandstone strata in the Lake District is evidence of Mid-Devonian movements.

Although palaeontological evidence is incomplete, it seems likely that the Millstone Grit succession is only partially developed in places in N.W. England, suggesting that movements occurred in the mid-Carboniferous.

Unconformity at the base of the Permian shows that important post-Carboniferous movements occurred, but on the whole Hercynian folding in N.W. England was comparatively gentle. The Lower Palaeozoic central block was arched into a dome with the Upper Palaeozoic strata dipping off it to the W. into the basin of the Cumberland Coalfield. This coalfield is cut by numerous faults, dominant directions being N.W. and N.E. These may have been initiated during the Hercynian movements but as many cut the Triassic their displacement is probably largely Tertiary.

On the E. side of the Lower Palaeozoic core the Carboniferous Limestone dips at about $30°$, into the New Red Sandstone basin of the Vale of Eden. This is bounded to the E. by the Pennine Fault, initiated during the Hercynian movements with upthrusting on the W. side. Differences in pebble contents between lower and higher Permian breccias have been taken as evidence of further movement of the Pennine Fault during the Permian but it has also been held that this is merely due to local accidents of denudation and sedimentation. Further movements of the Pennine Fault took place in the Tertiary.

Both N.E. and N.W. faults within the Lower Palaeozoic Block may also be Hercynian or at any rate have important Hercynian components. Moseley (1960 and 1964) has described N.W. faults near Ullswater which are thought to have an early phase of dextral or wrench-faulting and a later phase of normal faulting. The normal faulting is associated with mineralization and is probably post-Carboniferous, and it is possible that the wrench-faulting may also be Hercynian. Normal faulting of post-Carboniferous age also occurred in the Isle of Man (Simpson, 1963, 395).

Final uplift of the Lake District dome was almost certainly of Tertiary date. Triassic strata dip away from the Palaeozoic core over almost three-quarters of the periphery and Liassic strata are affected in the Carlisle Basin, where the synclinal amplitude exceeds 5000 ft. Moreover, the striking radial drainage, superimposed on the Caledonian trend-lines of the Lower Palaeozoic, can best be explained by Tertiary initiation on a thick Mesozoic cover. Bott (1974) attributes the uplift to the mass deficiency of a composite granite batholith underlying the Lake District and connecting exposed granites, etc. A negative gravity anomaly supports this view.

A powerful N.N.W. fracture brings down Triassic of the coastal strip against the western edge of the Lower Palaeozoic Block. The main movement of the Pennine Fault was also Tertiary, resulting in the downthrow of the New Red Sandstone of the Vale of Eden against the Carboniferous Limestone to the E. by not less than 4000 ft in reverse sense to the Hercynian displacement.

Gravity measurements in the Irish Sea (Bott, 1964) suggest that thick Carboniferous, New Red Sandstone and possibly post-Triassic rocks form the major "fill" of a number of deep basins (see Bott's Fig. 7) separated by horst-like areas of Lower Palaeozoic, of which the Isle of Man forms a part.

4.4 The Southern Uplands in Scotland and Ireland (Figs. 4.5, 4.6 and 4.7)

The Southern Uplands, the high ground between the populous Midland Valley of Scotland and the English Border, include a number of hills over 2500 ft (762 m) high, mostly with smooth, grass-covered slopes. More rugged scenery and the highest point, the Merrick, 2764 ft (842 m), occur in the S.W. where Lower Palaeozoic strata have been altered by granite. The Southern Uplands continue into Ireland, S. and S.W. of Belfast, where they form lower ground than in Scotland, in parts heavily drift covered; Tertiary intrusions (9.2), however, rise as mountainous country near the Irish Sea.

Some excursions are described by Mitchell and others (1960) and by Craig and Duff (1976).
The Southern Uplands are important in the history of structural geology for they include the unconformities of Upper Old Red Sandstone on Silurian at Siccar Point and near Jedburgh made famous by Hutton. Moreover, it was in this region that Lapworth discovered the zonal value of graptolites and from the stratigraphy went on to interpret the structure.

The succession is as follows :-
              Tertiary lavas (Ireland only)
              Triassic
              Permian
              Carboniferous
              Upper Old Red Sandstone (Scotland only)
              Lower Old Red Sandstone (Scotland only)
              Silurian
              Ordovician.

Since Lapworth demonstrated that 200 - 300 ft (61-91 m) of pelitic sediments near Moffat were deposited at the same time as 2000 - 3000 ft (610-915 m) of coarse sediments in the northern part of the region (and in the Girvan area), it has been realised that the Lower Palaeozoic strata of the Southern Uplands provide striking examples of facies-variation.

The Ordovician rocks are seen in the northern and central belts of the Southern Uplands and the Silurian in the southern belt (see below). Arenig strata are common in anticlines in the northern belt and consist of mudstones and cherts with lavas (including pillow-lavas), tuffs and agglomerates. There is a sedimentary break above the Arenig but its extent is disputed. Some authors hold that the Caradoc rests directly on the Arenig (cf. the Girvan district further N.W., 6.2), other authors that the Upper Llandeilo is present and others that the break is of only minor significance. It is, however, clear that in the central belt and the southern part of the northern belt the Upper Ordovician consists of a thin succession of dark pelites - the Glenkiln and Hartfell shales - while further N.W. it consists of thick greywackes, shales and conglomerates with occasional volcanics.

Ordovician strata, consisting mainly of greywackes, siltstones, mudstones, black shales and occasional black cherts and tuffs, form a strip up to 6 miles (4 km) wide in the N.W. of the Irish Southern Uplands and also appear in a number of anticlinal inliers within the Silurian further S. Fossils so far collected suggest that these rocks are of Bala age.

In the central belt of the Scottish Southern Uplands the Hartfell Shales are followed with no structural, but with marked palaeontological, break by the Birkhill shales which, together with thick flaggy greywackes and grits, belong to the Llandovery Series. The Wenlock Series occurs in the southern belt and consists of shales, greywackes, grits and conglomerates. Llandovery and Wenlock strata also occur in the Irish Southern Uplands. It is doubtful if Ludlow strata occur anywhere in the Southern Uplands. The Lower Palaeozoic rocks contain a number of mineral veins, notably in the Leadhills district where gold occurs.

The Old Red Sandstone occurs only in the Scottish Southern Uplands. The Lower division includes conglomerates, sandstones and volcanic rocks, mainly andesites; the Upper Old Red Sandstone consists of conglomerates and red sandstones in the lower part and paler sandstones with cornstones in the upper part.

Carboniferous strata occur in a few interior basins. In Scotland the succession is essentially similar to that of the Midland Valley (6.2), although in general thinner and less complete. The Kingscourt outlier in Ireland contains a Dinantian succession similar to that of the neighbouring parts of Central Ireland (6.7), followed by Namurian up to 2000 ft (610 m) thick consisting of pebble-beds, sandstones and shaley marine bands. Above come thin Coal Measures (Ammanian) consisting of argillaceous sandstones with subordinate shales and siltstones.

The largest outcrops of Permian rocks in Scotland occur in the Southern Uplands, consisting of breccias, red, dune-bedded sandstones and local basalt lavas. The Triassic is confined to a smaller area N. of the Solway, and includes red marls, shales and sandstones. In Ireland New Red Sandstone (Permian and Triassic) strata occur in the N.E. part of the area and in the Kingscourt Outlier. In the latter some 550 ft (168 m) of strata, consisting

of a basal conglomerate followed by grey shale with traces of plants and
marls with two thick gypsum/anhydrite beds, which are extensively mined, are
referred to the Permian. These are followed by about 1500 ft (457 m) of
sandstones, conglomerates, breccias and marls classified as Triassic.

A tongue of the main Tertiary basalt outcrop of Antrim overlaps onto the
Silurian S. of Lough Neagh and outliers continue as far S. as Kingscourt.
Large calc-alkaline plutons, which post-date the main Caledonian folding and
may be very late-Silurian or early-Devonian, occur in both parts of the
Southern Uplands, and in Ireland there are also Tertiary intrusions (9.2).

Tectonic Evolution

Structural events were :-

Caledonian	(Mid-Ordovician (Late-Silurian (Structures due to post-tectonic Caledonian intrusion (Mid-Devonian
Hercynian	(Pre-Namurian (Sudetic) (Post-Westphalian (Asturian)

Alpine (sensu lato).

The S.W. striking Southern Uplands Fault forms the N.W. boundary of the
region for 130 miles (208 km) from the North Sea coast near Dunbar to the
Irish Sea, with an en echelon shift towards the N., S. of Edinburgh. Much
remains to be learnt about this generally drift-covered fracture which
appears to be reversed, although this is not certain. The main movement,
with downthrow towards the N.W., was post-Lower Devonian and probably like
that along the Highland Boundary (3.2) mid-Devonian. Later movements (see
below) also occurred. The Fault, covered by Permian S.W. of Belfast,
continues across Ireland as far as Carrick-on-Shannon. Some authors consider
it dies out S.W. of Carrick, although Leake (1963) on geophysical evidence
suggests it continues at least as far as the neighbourhood of Galway.

Post-Arenig/Pre-Caradoc movements certainly took place in the Southern
Uplands, although the extent of the resulting stratigraphical break is
problematical. At least part of the Ordovician of the Northern Belt rose in
Ordovician times to the level of erosion for the Silurian sediments, derived
according to current determinations from the N.W., contain material of
Ordovician (and of metamorphic, perhaps Dalradian) provenance.

The main Caledonian phase, resulting in the production of dominantly N.E.
structures, was post-Wenlock and pre-Lower Devonian.

Following Peach and Horne (1899) it has long been accepted that the Lower
Palaeozoic strata form an anticlinorium in the N. and a synclinorium in the
S., made up of numerous isoclines. The structural units comprise a northern
belt, extending from the Lammermuir Hills to Loch Ryan and consisting of
Ordovician rocks, a broad central belt, from St. Abb's Head to the Mull of
Galloway, made up of the Llandovery Series with the Ordovician occurring in
long narrow anticlines and a southern belt from Jedburgh to Kirkcudbright
Bay occupied by the Wenlock strata.

The Lower Palaeozoic strata of the Southern Uplands do not, in general, show
cleavage comparable with that developed in strata of the same age in Wales.

Nevertheless, strong cleavage, amounting almost to schistosity, has been developed in narrow zones, for example in the Leadhills area.

As was shown by Lapworth (1878) and by Peach and Horne (1899) strike-faults are developed in association with the folds. Recent workers, including Craig and Walton (1959) and Walton (1961, 1963), have suggested that these play a major role and that the structural pattern is essentially that of north-westerly facing monoclines with strike-faults bringing up older strata to the N.W. Three phases of Caledonian folding have been recognised in several areas (Lindstrom, 1958; Kelling, 1961; Weir, 1968).

Post-tectonic plutonism produced significant structures. Thus, on both sides of the large dumb-bell-shaped Loch Doon Granite Complex the N.E. Caledonian structures swing through angles of $30°$ towards a N.-S. direction. Diversion of strike of $25°$ also occurs further S. on the north-western side of the Cairnsmore of Fleet mass. These are tectonic effects due to horizontal pressure accompanying magmatic intrusion. Dykes round the Cheviot granite, intruded into Lower Devonian lavas on the Border of England and Scotland, suggest radial tension accompanying intrusion.

The marked discordance between the Lower and the Upper divisions of the Old Red Sandstone in the Eyemouth district shows that the mid-Devonian folding, so strongly developed in the Midland Valley (6.2), affected at least part of the Southern Uplands.

Elsewhere the Upper Old Red Sandstone lies directly on the Lower Palaeozoic, for example at Hutton's classic Siccar Point unconformity.

Instability of the Southern Uplands during Carboniferous times is shown by the manner in which the Upper Carboniferous of the Sanquhar, Thornhill, Spango and Canonbie basins rests unconformably on the Lower Carboniferous or else overlaps onto the Lower Palaeozoic. In the Loch Ryan hollow no Lower Carboniferous is present, and the Upper Carboniferous directly overlies the Ordovician; S. of Dunbar and N. of the Solway only Lower Carboniferous occurs. The angular unconformity between Lower and Upper Carboniferous is, however, slight. A gravity survey of the Sanquhar Coalfield (McLean, 1961) suggests the presence of older Carboniferous strata, under known Coal Measures, lying in a pre-Westphalian basin trending N.W. and antedating the later Hercynian fold. It also suggests that proto-Hercynian movements took place along a north-easterly fault near Kirkconnel.

Post-Avonian movement took place in the Kingscourt trough in Ireland, as shown by the overstep of the Namurian.

Although the Lower Palaeozoic block of the Southern Uplands might be expected to be relatively resistant to N.-S. Hercynian compression, all the Carboniferous rocks of the region, in fact, show folding which can be attributed to this phase. Folding is N.W. in the Sanquhar Basin and N. in the Thornhill Basin. North-easterly folding occurs in the Carboniferous on the N. shores of the Solway. Such variations suggest that the effects of Hercynian compression were controlled by the different frameworks within which the Carboniferous rocks were deposited, as also applies to the Kingscourt trough. In the latter, and in the Thornhill, Canonbie and Loch Ryan basins, folding of the Carboniferous was pre-Permian, for the latter formation rests discordantly on the Upper Carboniferous.

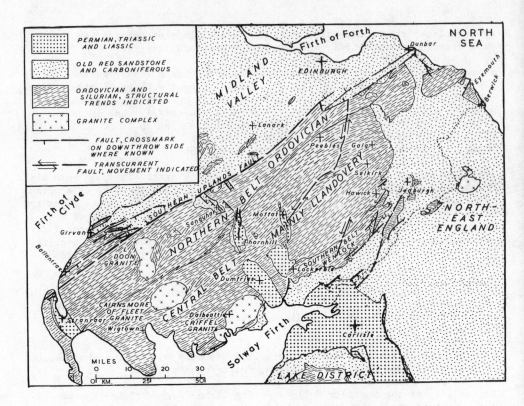

Fig. 4.6 STRUCTURAL MAP OF THE SOUTHERN UPLANDS

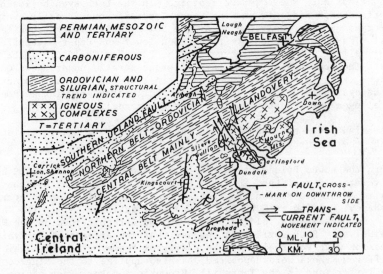

Fig. 4.7 STRUCTURAL MAP OF THE "SOUTHERN UPLANDS" IN IRELAND

Folding of likely Alpine date affected the New Red Sandstone of the Dumfries, Thornhill, Loch Ryan and Kingscourt Basins, and also probably the Sanquhar Coalfield. The Dumfries-Sanquhar depression could be the slightly sinuous continuation of the Vale of Eden (4.3).

Near the E. coast of Loch Ryan, a N.N.W. fault, which has probably a large Tertiary component, has been estimated by Kelling and Welsh (1970) to have a W.S.W. downthrow of about 5000 ft (1525 m) and to offset the Southern Uplands Fault. On the W. side of the Kingscourt trough a post-Triassic fault has a downthrow to the E. of 1200 to 2000 ft (365-610 m). Structures associated with Tertiary intrusions are described in 9.2.

4.5 South-East Ireland (Fig. 4.8)

South-East Ireland is a hilly to mountainous region consisting very dominantly of Lower Palaeozoic, invaded by the Leinster granite batholith, the largest in the British Isles, a mica-schist roof pendant of which forms the summit of the highest mountain, Lugnaquillia, 3039 ft (926 m). The block is bounded to the E. and S. by the sea and to the N. and W. mainly by the Lower Carboniferous of Central Ireland (6.7). Good coastal and hill sections occur, but the lower ground is frequently covered with glacial deposits.

The succession is as follows :-
>Carboniferous Limestone
>Old Red Sandstone
>Silurian
>Ordovician
>Cambrian
>Precambrian.

The only rocks of undoubted Precambrian age are those of the Rosslare Complex (Baker, 1955; Max, 1972) and of the Cullenstown Formation (Max & Dhonau, 1974) in the extreme S.E. of the region. The Rosslare Complex (for structural details see below) consists of gneisses and schists, invaded by the Rosslare Granite, poorly exposed because of glacial drift. The Cullenstown Formation, separated by Lower Carboniferous, consists of greywackes, quartzitic sandstone and phyllites. Like the Rosslare Complex it is undoubtedly pre-Ordovician and also has a different structure from rocks immediately to the N. correlated with the Bray Group which is now accepted as Cambrian.

At the N. end of the region the Bray Group forms a considerable outcrop S. of Dublin. The oldest part of this succession, the Devil's Glen Formation, consists of greywackes and slates and is overlain by the Bray Head Formation of quartzites (seen in the Sugar Loaf Mountain) and graded greywackes; all the rocks are turbidites. The Bray Group has been regarded as Precambrian, but microfossils from the Bray Head Formation indicate a Lower/Middle Cambrian age (Brück & Reeves, 1976). Across Dublin Bay the rocks of the Howth Peninsula are also turbidites of Cambrian age; they show large-scale slumping (van Lunsen & Max, 1975).

The Tramore Slates in the S. are Arenig and are overlain by Caradocian sediments and volcanics; in fact, the Llanvirn and Llandeilo appear to be missing. The widespread Ordovician outcrops on both sides of the Leinster Granite consist mainly of slates interbedded with thin, fine-grained, green and dark grey grits and cherts, with Caradocian limestones in places. The

Ashgill Series is unproved, and the limited Silurian outcrops, mainly slates, are probably all Llandovery.

The Old Red Sandstone occurs only on the S.W. margin of the region and overlies the Ordovician in a spectacular unconformity at Waterford.

Lower Carboniferous, dominantly of carbonate facies, occurs at Hook Head, E. of the Waterford ria, and in a graben between the two Precambrian formations near Rosslare.

Tectonic Evolution

Structural events were :-

	Precambrian
	(Mid-Ordovician
Caledonian	(?Late Ordovician
	(Late Silurian
	Hercynian

Field descriptions of parts of the region are given by Smyth et al., (1939) and by Bishop et al., (1948).

Precambrian events (Max, 1975) in the Rosslare Complex, which has a Lewisianoid aspect, can be summarised as follows.

EPISODE	ROCK TYPES
Fourth	Mylonites and folding
Third	Younger Basic Dykes
	Saltees Granite
	Older Basic Dykes
Second	Granodiorite Gneiss
	Intermediate Dykes
	St. Helen's Gabbro
	Fine-grained Amphibolite
First	Early Granodiorite Gneiss
	Grey Gneiss)
	'Initial Complex')
	Dark Gneiss)

The Carnsore Granite, which intrudes the metamorphics, is at least 550 m.y. old (Leutwein et al., 1972). The second episode of metamorphism is thought to have a minimum age of 1700-1600 m.y.

Breaks in the Ordovician succession show that mid-Lower Palaeozoic disturbances occurred, but the main Caledonian compressions were late Silurian, possibly persisting into the Devonian.

In a broad sense the Caledonian compression produced N.N.E. to N.E. anticlinoria in the Bray and Wexford areas with an intervening synclinorium in the Arklow district. South of Dublin the Bray Group has been thrust W. over

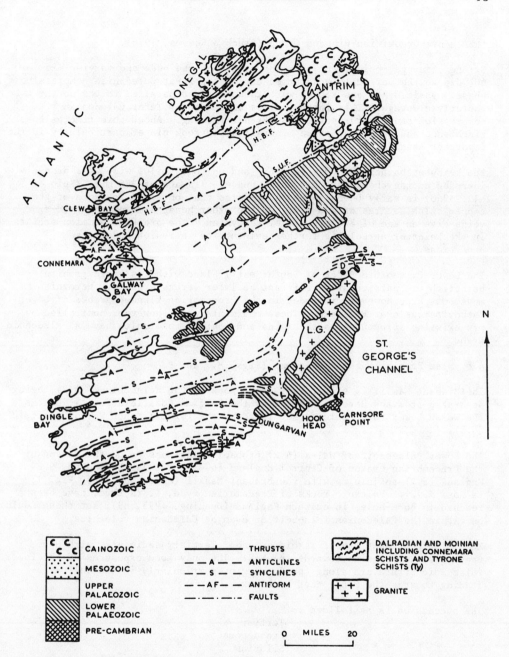

Fig. 4.8 STRUCTURAL MAP OF IRELAND
B. = Belfast; C. = Cork; D. = Dublin; R. = Rosslare;
L.G. = Leinster Granite; L.F. = Leannan Fault;
H.B.F. = Highland Boundary Fracture-Zone; S.U.F. = Southern Uplands Fault.

the Cambro-Ordovician Ribband Group (Brück & Reeves, 1976).

Strong cleavage, coincident with the strike of the beds and varying from a 60° dip to the vertical, was formed during the first Caledonian compressive phase. Basic dykes and stocks were then intruded parallel to the north-easterly cleavage. This intrusion was succeeded by later Caledonian compression causing thrusting and further folding. About this time galena, zincblende and copper sulphide mineralization took place, particularly in the important Avoca mining field.

The Leinster batholith, 70 miles long and up to 12 miles wide, was probably intruded during the later Caledonian phase. Its age of 386 m.y. (Kulp et al., 1960) is early Devonian and post-dates that obtained for some of the copper sulphide ores of 420 m.y. On the other hand, some of the lead-zinc veins give an age of 280 m.y., i.e. Hercynian. The presence of mica-schists in the Leinster aureole is further evidence of intrusion under stress conditions.

The Lower Palaeozoic Block of South-East Ireland with its large granite batholith was particularly resistant to later stress systems. Hercynian movements are, however, recorded in the Precambrian flanked graben of Lower Carboniferous near Rosslare. These movements were, however, controlled by pre-existing structures as the folds and most of the faults have a caledonoid trend.

4.6 The Brabant Massif (Belgium) (Figs. 8.7 and 8.8)

The Brabant Massif is structurally analogous to the Welsh Block (4.2), being separated from the Hercynian front by the Belgian Coalfield (5.11) just as the Welsh Lower Palaeozoic is separated by the (wider) South Wales Coalfield (6.5).

The Lower Palaeozoic of Wales (4.2) probably continues round the N. end of the Precambrian craton of Central England (6.4) and under London and S.E. England (8.8) to join up with the Brabant Massif (Anderson, 1965, Fig. 1; Le Bas, 1972). Volcanic rocks of Precambrian type, it is true, have been reached in bore-holes in eastern England (Dunning, 1975, 89), but these could be within the Caledonian fold-belt or even be Caledonian volcanics.

Much of the Brabant Massif is hidden under Tertiary sediments, with Upper Cretaceous intervening towards the S.E. The only good exposures of the Lower Palaeozoic strata are along the S. margin, particularly in the valleys leading towards the Meuse near Gembloux.

The succession is as follows :-

>Tertiary
>Cretaceous
>Silurian
>Ordovician
>Cambrian.

The base of the Cambrian is not seen; the succession, which consists of quartzites and phyllitic pelites, ranges up to the Tremadocian, which contains manganiferous beds (Fourmarier, 1954, 96).

Arenig fossils have not been found but there may, nevertheless, be a complete Ordovician succession consisting mainly of cleaved phyllites, siltstones and fine sandstones; Caradoc is well-developed in the Gembloux district. The Silurian, which ranges up to the Ludlow, is also dominantly pelitic with some sandstones and with calc-dolomite pelites and siltstones in the Wenlock. Tuffs occur in the Caradoc and tuffs and rhyolite lavas in the Llandovery.

The Cretaceous and Tertiary are similar to those of W. Belgium and S. Holland, although much thinner.

Tectonic Evolution

Structural events were :-
         Caledonian - Post-Silurian/pre M. Devonian
         Hercynian - Post-Westphalian (Asturian)

            Tertiary - faulting
            Plio-Pleistocene - uplift

Caledonian folding formed fairly tight folds, often overturned towards the S., with axial-plane cleavage in the finer beds. Folding and cleavage affect strata up to the Ludlow (Fourmarier, 1954, 27) which are unconformably covered by M. Devonian; this contrasts with the probably earlier main Caledonian movements in the Ardennes (5.11).

The Massif largely resisted Hercynian compression but the curvature of the strike, concave N. and parallel to that of the Belgian Coalfield, is of Hercynian date. Local fan arrangement of Caledonian cleavage has been ascribed to Hercynian refolding. The buried N.E. end of the Massif is cut by N.W. faults of the Tertiary Lower Rhine graben system (8.2).

Uplift during the Plio-Pleistocene was accompanied by downcutting of the present valleys.

REFERENCES

Anderson, J.G.C. 1965. The Precambrian of the British Isles. The Precambrian, vol. 2. (Ed. K. Rankama). Interscience Publishers, New York.
Baker, J.W. 1955. Precambrian Rocks in County Wexford. Geol. Mag., 92, 63.
Beavon, R.V. 1963. The Succession and Structure East of the Glaslyn River, North Wales. Quart. Journ. Geol. Soc. London., 119, 479-512.
Bishop, D.W. et al. 1948. The Geology of Eastern Ireland. Int. geol. Congr., London.
Bott, M.H.P. 1964. Gravity Measurements in the North-Eastern Part of the Irish Sea. Quart. Journ. Geol. Soc., London., 120, 309.
Bott, M.H.P. 1974. The Geological Interpretation of a Gravity Survey of the English Lake District and the Vale of Eden. Journ. Geol. Soc., 130, 309-331.
Brindley, D.W. et al. 1948. The Geology of Eastern Ireland. Int. Geol. Congr., London.
Brück, P.M. & Reeves, T.J. 1976. Stratigraphy, Sedimentology and Structure of the Bray Group in County Wicklow and South County Dublin. P.R.I.A., 76B, 53-77.
Craig, G.Y. & Duff, P.McL.D. 1976. The Geology of the Lothians and South-East Scotland. Edinburgh.

Craig, G.Y. & Walton, E.K. 1959. Sequence and Structure in the Silurian Rocks in Kirkcudbrightshire. Geol. Mag., 96, 209.
Dean, W.T. 1968. Geological Itineraries in South Shropshire. Geol. Assoc. Guide No. 27.
Dunning, F.W. 1975. Precambrian craton of Central England and the Welsh Borders. Sp. Rep. No. 6 The Precambrian, 83-94, Geol. Soc. London.
Fourmarier, P. (Ed.). 1954. Prodrome d'Une Description Géologique de la Belgique. Soc. Géol. de Belgique. Liège.
James, J.G. 1956. The Structure and Stratigraphy of Part of the Precambrian Outcrop between Church Stretton and Linley, Shropshire. Quart. Journ. Geol. Soc., London, 112, 315.
Kelling, G. 1961. The Stratigraphy and Structure of the Ordovician Rocks of the Rhinns of Galway. Quart. Journ. Geol. Soc., London, 117, 37.
Kelling, G. & Welsh, W. 1970. The Loch Ryan Fault. Scot. J. Geol., 6, 266-271.
Kulp, J.L. et al. 1960. Potassium-argon and rubidium-strontium ages of granites of Britain and Ireland. Nature, 185, 495.
Lapworth, C. 1878. The Moffat Series. Quart. Journ. Geol. Soc., London, 34, 240.
Leake, B.E. 1963. The Location of the Southern Upland Fault in Central Ireland. Geol. Mag., 100, 420.
Le Bas, M.J. 1972. Caledonian Igneous Rocks beneath Central and Eastern England. Proc. York. Geol. Soc., 39, 71-86.
Leutwein, F. et al. 1972. The Age of the Carnsore Granodiorite. Bull. geol. Surv. Ireland, 1, 303-309.
Lindstrom, M. 1958. Different Phases of Tectonic Deformation in the Rhinns of Galway. Nature, 182, 48.
Lunsen, H.A. van & Max, M.D. 1975. The geology of Howth and Ireland's Eye, Co. Dublin. Liverpool and Manch. Geol. J., 10, 35-58.
McLean, A.C. 1961. A Gravity Survey of the Sanquhar Coalfield. Proc. Royal Soc. Edinburgh, 68, 112.
Max, M.D. 1972. Two Pre-Cambrian Complexes in Ireland. Bull. geol. Surv. Ireland, 1, 99-105.
Max, M.D. 1975. Precambrian Rocks of South-East Ireland. Sp. Rep. 6 The Precambrian, 97-101, Geol. Soc. London.
Max, M.D. & Dhonau, N.B. 1974. The Cullenstown Formation: Late Precambrian Sediments in South-East Ireland. Bull. geol. Surv. Ireland, 1, 447-458.
Mitchell, G.H. 1970. The Lake District. Geol. Assoc. Guide No. 2.
Mitchell, G.H. & others. 1972. Excursion to the Northern Lake District. Proc. Geol. Assoc., 83, 443-470.
Mitchell, G.H., Walton, E.K. & Grant, D. (Eds.). 1960. Edinburgh Geology. An Excursion Guide. Oliver and Boyd, Edinburgh.
Moorbath, S. & Shackleton, R.M. 1966. Isotopic ages from the Precambrian Mona Complex of Anglesey, North Wales (Great Britain). Earth Planet. Sci. Lett., 10, 113-117.
Moseley, F. 1960. The Succession and Structure of the Borrowdale Volcanic Rocks, S.E. of Ullswater. Quart. Journ. Geol. Soc., London, 116, 55.
Moseley, F. 1964. The Succession and Structure of the Borrowdale Volcanic Rocks, N.W. of Ullswater. Liverpool and Manch. Geol. J., 4, 127.
Moseley, F. 1975. Structural Relations between the Skiddaw Slates and the Borrowdale Volcanics. Cumb. Geol. Soc., 3, 127-145.
Nichols, R.A.H. 1968. Structural Observations on the Carboniferous Limestone of the North Wales Coast. Geol. Mag., 105, 216-230.
Peach, B.N. & Horne, J. 1899. The Silurian Rocks of Britain. vol. I, Scotland, M.G.S.

Rast, B.R. 1961. Mid-Ordovician structures in South-Western Snowdonia. Liverpool and Manch. Geol. J., 2, 645.

Shackleton, R.M. 1954. The structure and succession of Anglesey and the Lleyn Peninsula. Adv. Sci., 11, 106.

Shackleton, R.M. 1958. The structural evolution of North Wales. Liverpool and Manch. Geol. J., 1, 261.

Simpson, A. 1963. The Stratigraphy and Tectonics of the Manx Slate Series, Isle of Man. Quart. Journ. Geol. Soc., London, 119, 367.

Smyth, L.B. et al. 1939. The Geology of South-East Ireland, together with parts of Limerick, Clare and Galway. Proc. Geol. Assoc., 50, 287-351.

Walton, E.K. 1961. Some Aspects of the Succession and Structure in the Lower Palaeozoic Rocks of the Southern Uplands of Scotland. Sonderdruck aus der Geologischen Rundschau Band, 50, 63-77.

Walton, E.K. 1963. In: The British Caledonides. (Eds. M.R.W. Johnson and F.H. Stewart). Oliver and Boyd, Edinburgh.

Weir, J.A. 1968. Structural History of the Silurian Rocks of the Coast W. of Gatehouse, Kirkcudbrightshire. Scot. J. Geol., 4, 31-52.

Williams, D. & Ramsay, J.G. 1959. Geology of Some Classic British Areas: Snowdonia. Geol. Assoc. Guide No. 28.

Williamson, J.A. 1967. Field Meeting at Great Langdale, Westmoreland. Proc. Geol. Assoc., 78, 489-491.

Chapter 5
HERCYNIAN FOLD - BELTS
(LARGELY METAMORPHIC)

5.1   The Massif Central and the Montagne Noire - Introduction

The Massif Central, the picturesque, dissected plateau which occupies one-sixth of France, is the spring-line for the Variscan (5.4 - 5.6) and Armorican (5.7 - 5.10) Arcs, the main Hercynian fold chains of Western Europe which extend outwards to Germany and Ireland respectively. The name Variscides is, however, given by some authors to the whole fold-system. Most of the Massif consists of highly folded metasediments (some of high grade) and granites of pre-Stephanian age. This Hercynian block now forms an uplifted peneplain around the 1000 m level. Lower areas, such as the Bas Limousin and the Hauts de Lyon are due to tilting towards the N.W. and differential uplift. Conversely, in the S.E., granite mountains rise to over 1700 m. On the eastern side of the Massif the crystalline rocks project northwards as the horst of the Morvan, rising W. of the Saône-Rhône Graben (8.6). On the S. side the Montagne Noire (5.3), although an extension of the Massif, is also related to the Pyrenees (7.16 - 7.21).

Owing to Plio-Pleistocene uplift, the rejuvenated rivers of the Massif have cut picturesque, wooded gorges in the metasediments and granites. The most spectacular features of the scenery are the recent volcanoes (9.3); the highest point, Mont Dore, 1885 m, is a large Pliocene volcano.

All major formations from the late Precambrian onwards, apart from the Cretaceous, are represented in the Massif Central and the Montagne Noire. Except in a few areas, however, of which the Montagne Noire and the Morvan are the most extensive, all pre-Stephanian strata are metamorphosed, both regionally and more locally, by granites which range from Devonian to Westphalian. The Stephanian forms several workable coal basins, some overlain by Permian; the most important is the St. Etienne field, the type-locality for the Stephanian.

Jurassic limestone, almost completely enclosed by metamorphics, make up Les Causses in the S. Permian, Triassic and Jurassic rock frame much of the Massif. Lower Tertiary sediments occur in grabens and basins. The upper Tertiary and Quaternary, apart from the glacial deposits, are almost entirely volcanic (9.3).

The deeply-incised valleys provide good sections;   excursions are described by Roques et al. (1954) and by Peterlongo (1972).

The main structure-building events were:-

Cadomian

Hercynian   (Pre-Stephanian, mainly Westphalian/Asturian
               (Post-Stephanian/pre-Permian - slight folding

Alpine   (Eocene - faulting; basin and graben formation
               (Early Miocene - faulting; accentuation of graben
               (Plio-Pleistocene - differential uplift

5.2 The Massif Central (Fig. 5.1)

The western third of the Massif is separated from the rest by the Sillon Houiller, a fault-zone which has been traced in a N.10°E. direction for 250 km. This originated as a transcurrent fault along which the E. side is thought to have moved N. between 40 and 50 km. It is often interpreted as the break between the Variscan and Armorican Arcs. The separation has been shown, however, not to be as sharp as once thought, as structures in the N.W. (Armorican) direction are found on the E. side and vice versa. The fault-zone also seems to bring together parts of the crust of contrasting thickness. Seismic studies indicate a crustal thickness of 23 km to the E. and of 30 km to the W. (Chenevoy, 1974, 164). Hercynian structural trends are also confused by their interaction with older trends, particularly in the N.

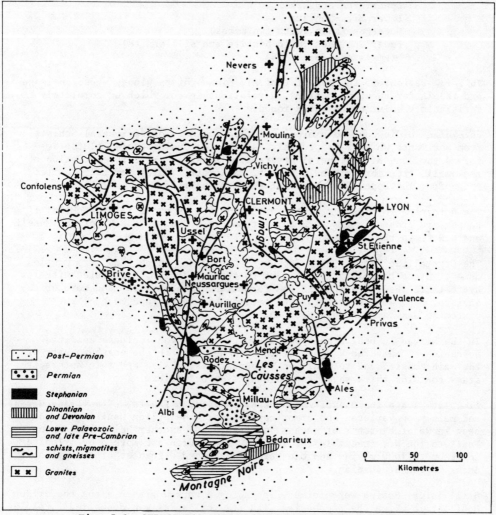

Fig. 5.1 STRUCTURAL MAP OF THE MASSIF CENTRAL OF FRANCE
(partly after Chenevoy 1974)

The Massif has been divided, on the basis of metamorphic studies, into the following units:-

>An ancient kernel or zone lemovico-arverue
>
>Zone of Peripheral Schists (W., S. and S.E.)
>
>Zone of Peripheral Crystalline Massifs.

The ancient kernel consists of metapsammites and metapelites with metavolcanics and some small masses of ultrabasics, all thought to be Precambrian. No stratigraphical succession has been made out, and a metamorphic rather than structural order has been established. Pre-Hercynian metamorphism has, in fact, developed the following zones.

>Upper Mica-Schists
>Lower Mica-Schists
>Upper Gneisses (with two micas)
>Lower Gneisses (with biotite and sillimanite)
>Migmatites   (Fig. 5.9)

In large regions of the Massif Central, mica-schists plunge under gneisses and the latter under migmatites, a relationship for which no completely satisfactory solution has been found.

Hercynian folding and metamorphism formed the Zone of Peripheral Schists from sediments which were probably Palaeozoic, and Hercynian plutonism formed the Zone of Peripheral Crystalline Massifs. With the exception of the small, Precambrian Mendic Massif, radiometric dates for the granites range from the Upper Devonian to the Namurian/Visean.

The Hercynian folding was of the anticlinorium type and it is doubtful if nappes are present. The structural trend is N.N.W. in the W. of the Massif and N.N.E. in the E.

The Crystalline Massifs often occur in anticlines which thus form mantled domes. Steep thermal gradients due to this plutonism, associated with low hydrostatic load, resulted in the formation of andalusite- and cordierite-schists. The main structural and metamorphic events and much of the plutonism was Namurian/Westphalian.

In the Morvan an unaltered Upper Devonian succession includes spilitic volcanics. Unaltered Visean occurs both here and in the northern part of the main massif, and in addition to sandstones, shales and thin anthracite seams contains thick rhyolite and dacite pyroclastics.

At a late stage the Massif was broken up by large faults, often transcurrent and mainly on Variscan and Armorican trends; the Sillon Houiller is the best known of a number of such major fractures, another being the Argentat Fault to the W., traceable for 160 km. The faults are marked by mylonite zones up to 1000 m wide and many are mineralised. Pebbles of the mylonite occur in the Stephanian.

Still later, basins were formed, including a narrow graben along the Sillon Houiller in which the Stephanian with its coal seams was deposited. Folding accentuated these basins before the Permian. Further transcurrent faulting along the Sillon Houiller produced complex structures in the Coal Measures.

During the Mesozoic the Massif formed a structurally stable block mostly above the level of sedimentation. Before the Jurassic the Massif became separated, probably initially by buckling and faulting, from the Variscan and Armorican Arcs by the Dijon and Poitiers gaps respectively. To the S. a great bay, probably also of tectonic origin, became flooded by an epicontinental Jurassic sea in which were deposited the thick limestones of Les Causses.

In the early Tertiary faulting along dominantly, although not exclusive, N.-S. lines and warping formed large Oligocene basins and grabens within the Massif, the most important being those of La Limagne, Roanne, Montbrison and Le Puy. The W. boundary fault of La Limagne, next which the Oligocene reaches its greatest thickness, has a downthrow of more than 2500 m N. of Clermont Ferrand. Complementary arching of the positive zones also occurred. Major faulting, much of the step type, took place between the eastern edge of the Massif and the Saône/Rhône Graben (8.6). The Jurassic limestone of Les Causses was faulted along N.-S., and W.N.W. and E.N.E. lines.

From the beginning of the Pliocene uplift occurred of an uneven character. These movements went on through the Pleistocene and continued to the present; there have been a number of fairly minor earthquakes, the most powerful in 1490 (Pelletier, 1969). Many of the Tertiary and Quaternary deformations had an important influence on the Tertiary and Quaternary vulcanism (9.3).

5.3   La Montagne Noire (Fig. 5.1)

Hercynian metamorphism did not occur in the Montagne Noire, but, on the other hand, Hercynian movements resulted in the formation of true nappe-structures. The Agout Dome in the centre of the Montagne Noire, consisting of migmatites and granites, is surrounded by a sedimentary succession ranging from late Precambrian (Brioverian) to Dinantian. Brioverian pelites are altered to slates or low-grade phyllites, but otherwise the sediments are unaltered and sufficiently fossiliferous for the Palaeozoic stratigraphy to be established. The absence of the Llandeilo Series and a difference in fold-style between the Cambrian and the higher Palaeozoic caused Gèze (1960) to postulate an early Caledonian orogeny in the Montagne Noire. This has been disputed by other authors who hold that there were at most epeirogenic movements; Rutten (1969) suggests that the difference in fold-style may be due to decollement above the Lower Ordovician as in the Pyrenees.

Nappes, marked by reversals of established Palaeozoic successions, were formed between the Dinantian and the Stephanian. The Mont Peyroux Nappe, on the S. flank of the central dome, consists of Ordovician-Devonian thrust over Lower Carboniferous. The nappes are themselves folded and their hinges are not seen. There is thus debate whether their roots lie to the N. or S. Rutten (1969, 165) has postulated a northerly origin by gliding from the rising Massif Central.

5.4   The Variscan Arc - Introduction

Three major outcrops of the Variscan Arc to be considered are the Vosges and the Schwarzwald (Black Forest) (5.5), and the Rheinisches Schiefergebirge

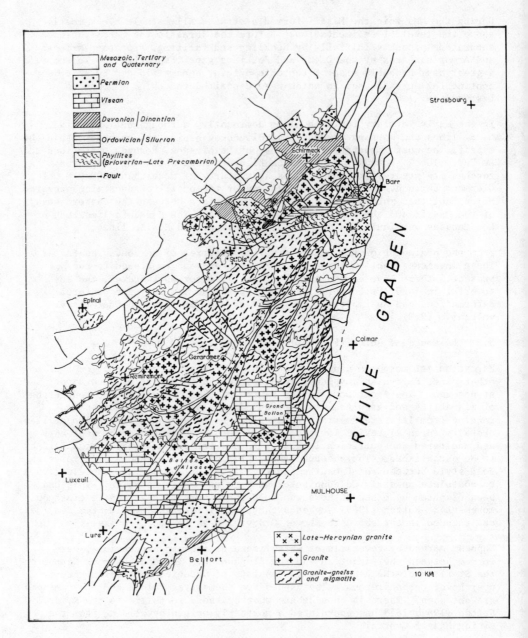

Fig. 5.2   STRUCTURAL MAP OF THE VOSGES, FRANCE

(Rhine Slate Mountains) (5.6). Between the Massif Central and the Vosges schist and granite are seen in the Serre inlier N. of Dole. The Vosges and the Schwarzwald are horsts separated by the Rhine graben (8.5). The graben and older, mainly Triassic strata on its flanks hide the Arc between these horsts and the Schiefergebirge, although granite is uplifted on the E. edge of the graben at Heidelburg.

The Arc is made up mainly of Silurian-Upper Carboniferous strata with limited outcrops of Permian and Triassic strata within the blocks and forming large outcrops on their margins.

Small basins of Lower Tertiary sediments occur in the Schiefergebirge where there are considerable outcrops of Tertiary/Quaternary volcanics (9.4).

The Vosges and the Schwarzwald, like the Massif Central, are largely made up of metasediments and granites; the latter can be shown to be intruded into Devonian and Carboniferous strata. Late Precambrian metamorphics form cores for both blocks. In the Schiefergebirge pelitic sediments are altered only to slates with granites absent. Further N.E., however, in the Harz Mountains, granites reappear.

Structure-building events are:-

>   Late Precambrian - Cadomian
>   Hercynian
>   Eocene - faulting; basin and graben formation
>   Early Miocene - faulting; accentuation of grabens
>   Plio-Pleistocene - differential uplift

5.5   The Vosges and the Schwarzwald (Fig. 5.2)

The Vosges are rounded hills rising to 1423 m; the higher parts have been glaciated. There are numerous craggy outcrops of fresh rock, and good sections along some of the rivers, but the thick woods, even on the higher slopes, and the widespread, low-level glacial outwash make structures difficult to follow. Some excursions are described by Wallace and Laurentiaux (1973).

The greater part of the Vosges consists of gneisses, migmatites, metapelites, metapsammites and granites. Some of the metamorphics have been dated as probably Precambrian and correlated with the Zone lemovico-arverne of the Massif Central (5.2). Nevertheless, the structural trend is strongly N.N.E., i.e. Variscan. The granites, on the other hand, are Hercynian; the well-known Barr-Andlau granite contact metamorphoses Silurian and others Devonian or Lower Carboniferous. Both marginally and within the block the Hercynian or older rocks and structures are overlain by Triassic conglomerates and grits; Permian occurs in the S. North-north-easterly steep faults form the E. margin of the block against the Rhine graben; fractures in the same direction occur within the block, including the Grande Faille des Vosges which passes E. of Gérardmer and has been traced for 50 km; it probably continues for double that distance as far S. as the Jura. A number of faults with general easterly trend also occur. Many of the faults are Tertiary but they may have moved along older fractures; springs up to 53°C show that some of the faults are deep-seated.

Both topographically and geologically the Schwarzwald is similar to the

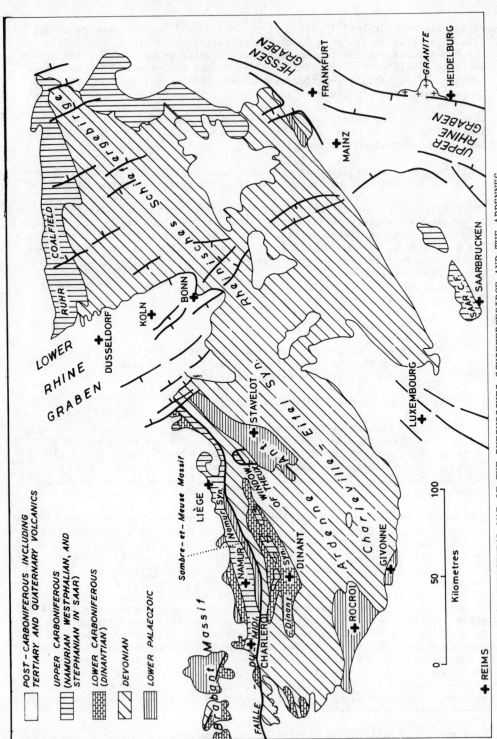

Fig. 5.3 STRUCTURAL MAP OF THE RHEINISCHES SCHIEFERGEBIRGE AND THE ARDENNES

Vosges. Rounded hills, mostly wooded, rise to 1493 m.

Parts, at least, of the metamorphics are considered to be late Precambrian, but the structural trend is Variscan; all the granites are probably Hercynian. The W. margin against the Rhine graben is sharply defined by N.N.E. faults. Some powerful E.-W. fractures cut the block, including those which let down a long, narrow strip of Lower Carboniferous at Schonau and another forming the southern margin of the block.

5.6 The Rheinisches Schiefergebirge and the Ruhr Coalfield (Fig. 5.3)

The Rhine Slate Mountains, in N.W. Germany, rise to just under 900 m athwart the Rhine from Mainz to Bonn. Excellent sections occur on the sides of the Rhine and its tributary valleys, deeply incised during Plio/Pleistocene uplift. Excursions are described by Tilman et al. (1938).

The mountains are made up largely of Lower/Middle Devonian clastic sediments. The Devonian opened in the Lahn and Dill areas with trachyte and spilite lavas (often of pillow type) associated with sills of the same composition. The pelitic sediments have been metamorphosed to slates; the sandy sediments are interpreted as turbidites.

The Lower Carboniferous, unlike that of the Ardennes (5.11), consists mainly of clastic sediments, the Visean, in particular, being of Culm facies; volcanics also occur. The Upper Carboniferous occurs on the N. margin where it passes downwards into the Ruhr Coalfield, occupying a tectonic depression bordering the Chain and characterised by structures similar to those of the Belgian Coalfield (5.11).

Tectonically, the Schiefergebirge consists of a number of tightly-folded anticlinoria and synclinoria with overthrusts. Broad zones, in which the over-folding and overthrusting are either northwards or southwards, are separated by zones in which a subvertical fan-like regime predominates. Some overthrusts bring up narrow strips of Lower Palaeozoic.

In the N. the Lower Rhine Graben (8.5) forms an embayment in the Chain bordered and cut by a number of N.W. Tertiary faults. This fault-system extends through the Schiefergebirge and forms a link with the Upper Rhine Graben (8.5). A branch of the latter continues N. through the Chain as the Hessen Graben.

Uneven uplift, involving further faulting, occurred in the Plio-Pleistocene. The faulting largely controlled widespread Tertiary/Quaternary vulcanicity (9.4).

5.7 The Armorican Arc - Introduction

Three large, separated sectors of the Arc are seen above the present level of erosion, i.e. in Armorica (Brittany, La Vendee, parts of Normandy and the Channel Islands (Iles Anglo-Normandes)), in South-West England and in Southern Ireland. Although the structures and rocks of the Arc have been shown to extend some distance out from the coastlines, they are covered largely by younger formations forming the floor of the English Channel and the southern Celtic Sea (8.11).

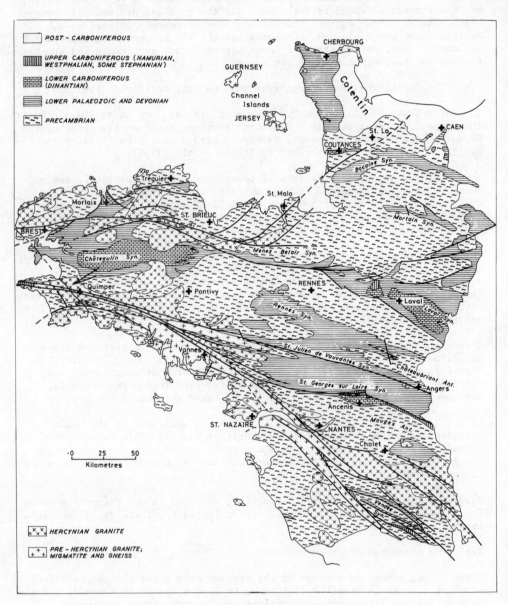

Fig. 5.4 STRUCTURAL MAP OF ARMORICA

The Arc shows a marked curvature from N.N.W. in the Western Massif Central (5.2) to W. by S. in Ireland.

The rocks of the Arc range from Precambrian to Stephanian, with the Late Precambrian particularly important in Armorica. The outcrops of the Arc are flanked by Permian and Mesozoic strata.

The main structural events were:-

>Early Precambrian
>Middle Precambrian
>Late Precambrian - Cadomian
>Pre-Namurian - Sudetic
>Post-Westphalian - Asturian
>Middle Tertiary
>Pliocene - uplift

5.8   The Armorican Arc - France and the Channel Islands (Fig. 5.4)

The Armorican Massif is an upland rather than a mountain area rising to some 400 m. The region escaped Pleistocene glaciation although permafrost effects are evident. Inland exposures of fresh rock are comparatively rare, but this is compensated by the excellent coastal sections, some along ria-like inlets. Some of these follow the Hercynian structural trends, as do many of the topographical features of the region. The Channel Islands are geologically part of the Cherbourg or Cotentin peninsula.

The Icart orthogneisses of Guernsey have been dated at 2620 m.y., and the name Icartian has been suggested for this event (Bishop et al., 1975, 104). Petrologically similar rocks near Cherbourg have given 2650 m.y. The Penteverian (Sarnian) Complex, consisting mainly of granitic gneisses folded along northerly axes, has been dated at 1000-1200 m.y. These gneisses are overlain unconformably near St. Brieuc by late Precambrian (Brioverian). Other metamorphic rocks in Southern Brittany, in the N.W. (Léon) and in the Channel Islands may also be Penteverian.

Much of Brittany is made up of a late Precambrian formation, the Brioverian or Vendian. The Brioverian is noteworthy for the occurrence of a glacogene, on the presence of which, it is generally correlated with glacogene-bearing late Precambrian in the Caledonides of Scandinavia, Scotland and Ireland. The Precambrian rocks appear in the major Armorican anticlinoria with Palaeozoic strata preserved in narrower major downfolds.

Parts of the Brioverian are highly folded and metamorphosed (up to the sillimanite grade) and have been intruded by Pre-Hercynian granite. On the other hand, parts are only slightly metamorphosed, e.g. E. of Brest; in addition to the glacogenes already mentioned, shales, graded greywackes and spilitic volcanics with pillow structure are present.

Just before and into the beginning of the Cambrian, the Precambrian strata of Armorica were strongly folded and in places metamorphosed by the Cadomian orogeny (Latin Cadomus = Caen); the accompanying granite intrusion gives a date of about 580 m.y. Brioverian sediments folded and slightly metamorphosed during the Cadomian orogeny occur in the Channel Islands, and on Jersey, Guernsey and Alderney were intruded late in the Cadomian orogeny by gabbro-diorite-granite igneous complexes. On Jersey the Brioverian metasediments are overlain by virtually unmetamorphosed andesitic and rhyolitic volcanic rocks of uncertain age.

After a long period of Post-Cadomian erosion, the Lower Palaeozoic/Devonian succession was laid down, consisting essentially of conglomerates, sandstones, quartzites and shales. The succession includes the well-known Grès Armoricain of Brittany, a beach facies deposit of quartzite up to 1000 m thick. Limestones come on in the Upper Devonian.

The Alderney Sandstone, and the Rozel Conglomerate of Jersey, post-date the late Cadomian igneous activity and are probably post-orogenic, molasse-type deposits of Cambrian age (Bishop et al., 1975, 106).

There was no Caledonian orogeny in Armorica, but breaks in the succession just outlined show that there was some uplift and subsidence. An early manifestation of the Hercynian orogeny is shown by an important break at the base of the Carboniferous (the Bretonic Phase). The main phase was probably Pre-Namurian (Sudetic) but it is a matter of opinion whether this was more important than the pre-Stephanian (Asturian) phase, for in places the earliest post-tectonic molasse deposits are Stephanian, in others Westphalian.

The Hercynian folds are open flexures arranged in long anticlinoria and synclinoria. In some cases the Palaeozoic cover folded while the basement responded by faulting. Near Cherbourg the Brioverian is cut by imbricated reversed faults which Graindor (1961) takes as evidence of decollement between this formation and the more highly folded Palaeozoic cover, a view disputed by Klein (1963). Migmatite injection and granite intrusion, largely along growing master fractures, were also spread over a considerable interval. Large sinistral, E.-W. wrench fractures occurred late in the orogeny. Deposition of the Stephanian clastics and thin limnic coals was followed by only slight folding but extensive faulting. Mid-Tertiary movements produced shallow folds and re-activation of older faults.

5.9 The Armorican Arc - South-West England (Fig. 5.5)

The Cornubian Peninsula, the part of England between the Bristol Channel and the English Channel W. of Exeter, is an upland rather than a mountain area with Devonian grit in Exmoor reaching 1705 ft (520 m), and the granite of Dartmoor 2039 ft (620 m). The region was virtually unaffected by glaciation, although permafrost effects are evident. Apart from the granites and their contact aureoles, inland exposures are scattered, especially on the 430 ft (130 m) platform which forms a wide coastal fringe. There are, however, many excellent cliff exposures, and river valleys, such as the Dart, incised by recent uplift into the 430 ft platform, provide good sections. Some excursions are described by Dearman (1959), Dearman et al. (1970), Holwill et al. (1969) and by Hall (1974).

By far the greater part of South-West England consists of Devonian and Carboniferous. Lower Palaeozoic is confined to a few small outcrops in the S., and the status of some metamorphic rocks identified as Precambrian is doubtful. The region is underlain by a huge E.N.E. orientated granite batholith, large cupolas of which form granite outcrops from Dartmoor to the Isles of Scilly. The batholith has been responsible for the formation of the well-known Cornish tin and other deposits, and pneumatolysis of the granite has formed valuable china clay deposits. Unconformable Permian forms the E. boundary and a few small outliers; there is a fairly large Oligocene basin.

In contrast with Armorica, there is very little evidence of Pre-Hercynian structural events. Such clues as there are lie entirely in masses of deformed and metamorphosed rocks thrust from the S. onto the Devonian at Start Point in the Lizard. The occurrence of thrusts is supported by geophysical results (Bott et al., 1958). The Start Point metamorphics consist of mica-schists and hornblende-schists from which a date of about 300 m.y. has been obtained (Dodson & Rex, 1971). This would be Caledonian (sensu lato), but later tectonisation could have "diluted" the date and the schists could be Cadomian brought northwards.

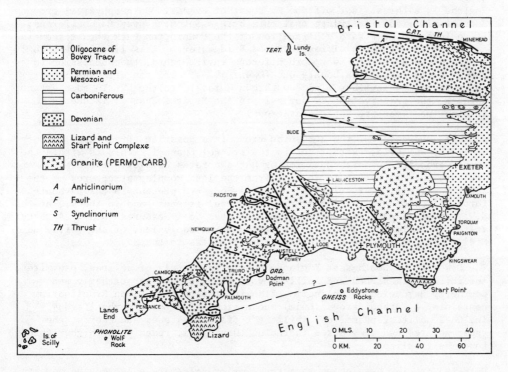

Fig. 5.5   STRUCTURES OF SOUTH-WEST ENGLAND
C.P.T. = Cannington Park Thrust

The Lizard Peninsula includes mica- and hornblende-schists invaded by serpentinised peridotite, gabbro and acid material altered to the Kennack Gneiss. The latter has yielded dates of 350-390 m.y. The older date is Caledonian (sensu stricto), but the younger could be Bretonic, i.e. Hercynian (sensu lato). The hornblende-schists have given 442 and 492 m.y. (Miller & Green, 1961), which again could be "diluted" Cadomian ages.

Between the Lizard rocks and the Devonian the Meneage Melange contains lenses which include Lower Palaeozoic strata interpreted as the remains of nappes. Subaqueous sliding has also been postulated (Lambert, 1964). Between Start Point and the Lizard, Normandy-type Ordovician quartzite, near Dodman (S. of St. Austell), has been thrust from the S. Some geologists believe that a major thrust separates Cornubia from Armorica.

A possible northern "front" for the complex Hercynian structures of S.W. England is a thrust just off the N. coast of Exmoor. On geophysical grounds Bott et al. (1958) postulate that this brings Devonian over Carboniferous. It could be the westward continuation of the Cannington Park thrust seen in an inlier surrounded by Triassic S. of Bridgwater. This line appears to separate two distinct Lower Carboniferous environments, the succession to the N. being a shelf carbonate one (Owen, 1974). Off the straight Exmoor coast fault movement has also downthrown Jurassic to the N. To some extent, however, the preservation of Jurassic in the Bristol Channel is due to folding (possible pre-Upper Cretaceous).

Another candidate for the Hercynian front is a possible thrust zone, mainly under the Jurassic, further out in the Bristol Channel. This is within the Lower Carboniferous carbonate facies but separates complex structures, including overthrusting to the S., from the less complex structures of the Welsh Coalfield. Thrusting and vertical beds are present in the Carboniferous Limestone of the inner Bristol Channel islands, and to the W., at the S.W. tip of Wales, Precambrian is pushed over Coal Measures (6.5). A thrust in this position could be a continuation or an equivalent of the Faille du Midi (5.11), which has also Dinantian carbonates to the S.

The Devonian of South-West England, consisting mainly of pelites, psammites, turbidites, limestones and spilitic volcanics, occurs in northerly and southerly Hercynian anticlinoria, flanking the Central Devon Synclinorium containing a Carboniferous (Culm) succession. The Lower Carboniferous includes pelites, psammites, spilitic tuffs, limestones and cherts, the Upper pelites and greywackes including turbidites, reaching no higher than Lower Westphalian.

Detailed studies have shown that these major folds are superimposed on complex polyphase structures (e.g. Dearman, 1970). On the S. margin of the main synclinorium Hobson and Sanderson (1975) have demonstrated the presence of early, major, southerly-facing, recumbent folds modified by later low-angled faults. Dearman (1969) has suggested that the phases of folding moved steadily northwards. Metamorphism accompanying the folding did not go beyond the production of slates (workable in places) and in some places more than one cleavage can be made out. The youngest strata affected by the folding are Lower Westphalian, and the granite batholith, dated at about 290 m.y., is certainly younger. The flood of debris transported northwards into the Pennant-facies of the South Wales, Bristol and Kent Coalfields in Upper Westphalian times suggests that the Cornubian folding could at least have commenced as early as the mid-Westphalian (Owen, 1971).

The Mesozoic history was one of uplift and subsidence, but important faulting took place in Tertiary times, probably after the Oligocene, as fractures cut this formation in the Bovey Tracey Basin near Torquay. Furthermore, the important Sticklepath Fault seems to form the E. margin of the Lundy Island granite in the Bristol Channel dated as Eocene. Many of the faults run on a N.N.W. direction and Dearman (1963) has shown that these have moved dextrally. If this movement is "put back" it has the effect of placing the granite cupolas on a more strictly E.-W. line.

Uplift took place in the Pliocene raising the 430 ft and other platforms. Earlier, Neogene uplifts could be represented by higher platforms (1000 ft, etc.).

5.10 The Armorican Arc - Southern Ireland (Fig. 5.6)

Much of the Armorican Arc in Southern Ireland forms rugged ground, culminating in Carrauntoohil, 3414 ft (1035 m), the highest point in Ireland, formed of hard, fine-grained Devonian sandstone. The whole region has undergone Pleistocene glaciation, which left many rock outcrops on the higher ground but covered the lower areas with thick drift deposits. There are good exposures, however, along the coast, which, in the S.W., is deeply indented by strike-controlled ria-inlets.

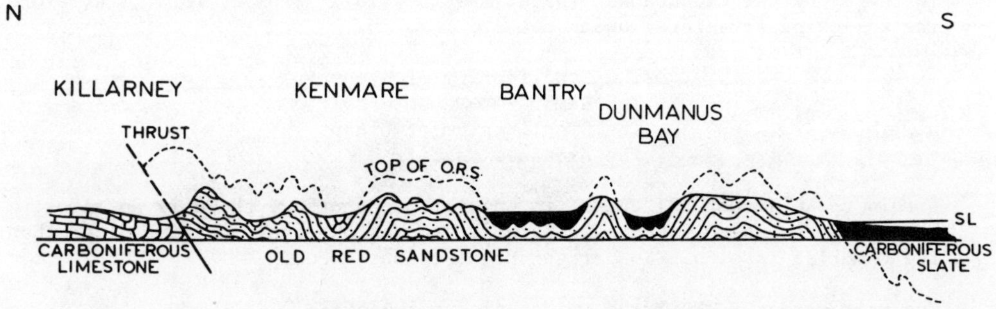

Fig. 5.6 DIAGRAMMATIC SECTION OF SOUTH-WEST IRELAND

Only highly folded Devonian and Carboniferous rocks are present. These are separated by the Hercynian front, extending from Dungarvan Harbour to Dingle Bay, from a structurally simpler region further N. where, moreover, the Devonian and Carboniferous sedimentary piles show rapid thinning. Near Killarney the front is marked by powerful, mainly reversed, faults (Walsh, 1968). Gill (in: Coe, 1962) states that further E. the dislocation is for the most part along a single thrust dipping S. at $45°$. Further E. still, towards Dungarvan, Philcox (1963) believes that the front splits up into echelon zones and that this "fanning out" may be related to Caledonoid influences.

The thick, dominantly sandy, Devonian succession of Southern Ireland appears in a number of complex, major anticlinoria. Downfolds preserve a Lower Carboniferous succession similar to that of Cornwall. The Devonian succession is up to 7000 m thick S. of Dingle Bay and the Lower Carboniferous over 2500 m thick. Pelites have been altered to slate. There is a complex fault pattern including E.N.E. thrusts, powerful wrench faults trending E., and a large number of fractures ranging from N.W. through N. to N.N.E. Gill (in: Coe, 1962) has noted how N.W. wrench movements accompanied the growth of the folds. Northerly faults in the mountains S. of Dingle Bay must have a large accumulative easterly downthrow which cancels out the persistent westerly plunge of the folds in this area (Walsh, 1968, 19).

5.11   The Ardennes and the Belgian Coalfield (Figs. 5.3 and 5.7)

The Ardennes proper and the Synclinorium of Dinant immediately to the N. are part of a highly folded Hercynian Chain crossing the gap between the Variscan and Hercynian Arcs.

To the N. the Hercynian front is strongly-defined by the overthrust of the Faille du Midi. As the name implies this forms the southerly boundary of the Belgian Coalfield. Although the Coalfield thus lies outside the main fold-belt it is so intimately connected with the latter that it is described in the present section.

The Coalfield, in fact, lies in the Synclinorium of Namur which is bounded to the N. by the Caledonian Brabant Massif (4.6). From N. to S., therefore, the following structural units occur:-

>   The Synclinorium of Namur
>   The Sambre-et-Meuse Massif
>   The Synclinorium of Dinant
>   The Ardennes Massif

Within these structural units, in one part or another, there is an almost complete succession from Lower Cambrian to Westphalian; nothing older than the Cambrian is seen. No Hercynian intrusions occur.

The main structure-building events are as follows:-

>   Caledonian (sensu lato)
>   Post-Westphalian - Asturian
>   Plio-Pleistocene - uplift

Topographically the Ardennes and their flanks form rolling, wooded hills rising to somewhat over the 500 m level. Exposures on these hills are scarce but there are excellent sections along the Meuse and along the deeply incised tributaries that flow into it from the S. Some excursions are described by Fourmarier (1950), by Wallace and Laurentiaux (1973) and by Waterlot et al. (1973).

The Devonian of the N. margin of the Synclinorium of Namur rests on the southerly-inclined southern edge of the Brabant Massif. At the N. margin of the Synclinorium the Lower Devonian is generally missing and at some places

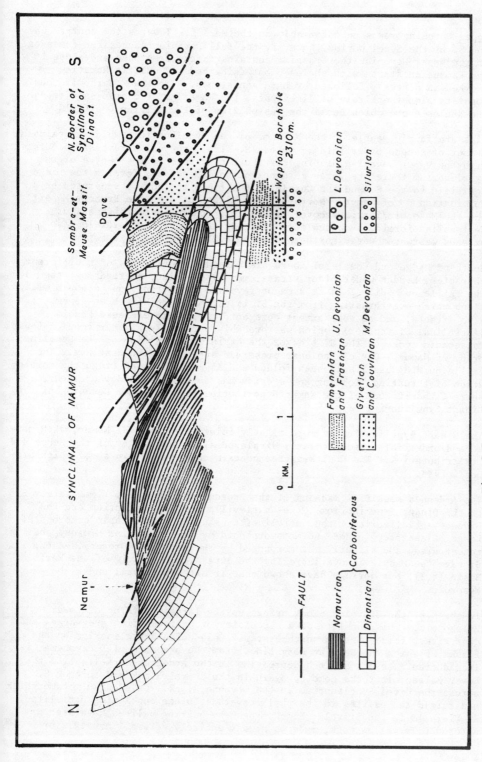

Fig. 5.7 SECTION OF SYNCLINAL OF NAMUR

Upper Devonian rests unconformably on the Massif. Towards the centre and the S. of the Synclinorium, however, the full Devonian succession comes on. In the Upper Devonian the Frasnian contains thick limestones and above it the Famennian starts with shales, containing beds of oolitic hematite (worked in places), followed by hard sandstones. The Lower Carboniferous consists almost entirely of limestones, the Namurian of coarse sandstones and shales above which comes the Westphalian of the Belgian Coalfield.

Dips are fairly gentle on the N. limb of the Synclinorium, which contains a number of second order folds. On the S. side, however, a reversal occurs with the Upper Devonian dipping off the Dinantian. In the centre of the Synclinorium there are large outcrops of Coal Measures, forming the Liège Coalfield to the E. and the Charleroi Coalfield to the W., separated by a culmination E. of Namur. North-east of Liège, between the Brabant Massif and the Mesozoic/Tertiary cover, there is a considerable thickness of Westphalian forming a concealed coalfield. A concealed coalfield also extends westwards under northern France (8.2).

South of Liège and Charleroi there are concealed coalfields of a different character, hidden under older strata, mainly Devonian, carried over them by the Faille du Midi. This is a major reversed fault-zone, up to 600 m wide, along which overthrusting from the S. is estimated to have been 15 km. It also affects, and to some extent cuts out, the Sambre-et-Meuse Massif (Anticlinal of Condroz), which was probably originally a wider zone of Lower Palaeozoic rocks on the S. side of the basin of deposition of the Synclinorium of Namur. One of the best preserved strips is seen near Huy. The Faille du Midi has itself been folded and 15 miles S.E. of Liège the combination of thrusting, folding and erosion has formed the Window of Theux, where Carboniferous of the Namur Synclinorium is seen under Devonian of the Dinant Synclinorium.

The Dinant Synclinorium is more tightly folded than that of Namur, with many second order folds spectacularly displayed along the valley of the Meuse and its tributaries. The Coal Measures contain only a few thin seams which are not worked.

The Ardennes Massif is made up of the Ardenne Anticlinorium (immediately S. of the Dinant Synclinorium), the Charleville-Eifel Synclinorium and the Givonne Anticlinorium; the anticlinoria bring up considerable outcrops of Lower Palaeozoic, folded and metamorphosed by the Caledonian Orogeny and retectonised and slightly metamorphosed by the Hercynian Orogeny. Along its S. edge the Massif is overlapped by the Jurassic on the rim of the Paris Basin (8.2), but borings have shown that it extends S. well under the Mesozoic.

In the N.E. the Stavolot Block brings up the largest outcrop of Lower Palaeozoic. This consists of a succession of pelites, fine greywackes and quartzites, folded into a north-easterly Caledonian anticlinorium on the flanks of which the Salmian Formation, from the presence of Dictyonema, is Tremadocian. Beyond a pitch depression in the Ardennes Anticlinorium the Lower Palaeozoic rocks come up again in the Rocroi Block and further S. across the Eifel Synclinorium in the Givonne Block. Caledonian metamorphism has raised the pelites to the phyllite grade, often containing ottrelite. At Vielsalm in the Stavolot Block manganese-bearing pelites have been altered to a yellow rock, made up almost entirely of small crystals of spessartite-garnet, extensively quarried as honestones. In the Brabant Massif (4.6) the main Caledonian movements came at the end of the Silurian, but in the Ardennes proper they may well have been late Ordovician (Taconic) as the youngest strata affected are Caradocian and the Lower Palaeozoic clearly underwent deep erosion before the deposition of the Gedinian. Hercynian movements resulted in the formation of the anticlinoria just described and of the intervening Eifel Synclinorium, marked by fairly open and symmetrical folds in the Devonian. These movements were accompanied by cleavage development in pelites, including, in some places, a second cleavage cutting the earlier Caledonian cleavage in the older rocks. There was a slight thermal Hercynian alteration to which is attributed a second generation ottrelite at Vielsalm (Michot, 1955). Furthermore, in the Poudingue de Fépin, the basal conglomerate of the Gedinian resting on the Cambrian at the N. edge of the Rocroi Massif in the Meuse Valley, minerals have grown from the cement into the pebbles due to light regional metamorphism.

The Mesozoic and most of the Tertiary history of the Ardennes was one of minor vertical movements, but at about the time of the transition from Pliocene to Pleistocene the region became an uplifted block.

5.12   The Iberian Peninsula (Fig. 5.8) - Introduction

Apart from Alpine Chains (7.15-7.23) and Tertiary basins, Spain and Portugal are made up entirely of Hercynian tectogenes and plutons. In the N., the Cantabrian and León-West Asturia Zones (5.13) strike southwards from the Biscay coast and then curve eastwards. The rest of Hercynian Iberia is made up of the Central Iberian (Galician/Castillian) Zone, followed to the S.W. by the Ossa/Morena and South Portuguese Zones (5.14). Much of the Central Iberian Zone has been uplifted to form the high, often dry, plateaus (mesetas), diversified by mountains (sierras). Although Precambrian strata have been thrust or folded up in places, Hercynian Spain and Portugal consist dominantly of Palaeozoic rocks, both metamorphic and unaltered, invaded by granite masses.

The chief structural events were:-

                Precambrian

                (Early Carboniferous - Bretonic   - mainly uplift
Hercynian   (Post-Westphalian   - Asturian
                (Post-Stephanian   - Saalian

                Early to Mid Tertiary   - faulting; basin and graben formation

                Plio-Pleistocene   - uplift

5.13   The Cantabrian and León-West Asturias Zones (Fig. 5.8)

The rocks of these zones form mountainous country in Northern Spain; several peaks, snow-covered for much of the year, rise to over 2000 m, culminating in the Picos de Europa, 2642 m. Above about 550 m the mountains were glaciated, and thick fluvioglacial deposits occur in the valleys. The higher parts are bare, as are the relatively dry southern slopes. The wet N. slopes are heavily wooded and drop to a narrow, rather irregular, coastal platform bordered by fine cliff sections from San Vicente de la Barquera

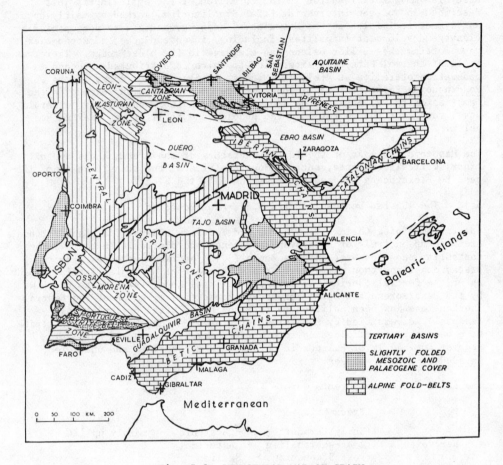

Fig. 5.8 STRUCTURAL MAP OF SPAIN

westwards. Excursions in the region are described by Wallace (1972b).

The Cantabrian Chain is characterised by a shallow-water succession, ranging from late Precambrian to Upper Carboniferous, which has undergone several Hercynian phases of folding but has escaped metamorphism. The clastic Mora formation is Precambrian as it is followed unconformably by the early Cambrian Herreria formation. The Cambrian and Ordovician consist of sandstones, quartzites, dolomites and nodular limestones (griotte). The Silurian is much thinner and is made up largely of shales in which the zonal graptolites are found. Much of the Devonian consists of carbonates.

The Dinantian is represented mainly by carbonates, shales and radiolarian cherts. The carbonates include the nodular Alba Griotte, almost indistinguishable from the Cambrian Lancara Griotte apart from its fauna. The Caliza de Montana formation, of Namurian age, contains massive limestones, conspicuous in the mountain and coastal scenery. At the top of this succession a terrigenous sequence of sandstones and shales appears. Similar strata occur in the Westphalian, in which there are thick clastics in the S., but a more varied succession of conglomerates, sandstones, siltstones, carbonaceous shales, coals and fireclays further N. The coals are worked in the Central Asturian Coalfield, S. and S.E. of Oviedo. These strata are highly tectonised, and steep dips complicate mining.

The Westphalian is followed unconformably (de Sitter, in: Coe, 1962; Wallace, 1972a) by Stephanian Coal Measures consisting of conglomerates, sandstones, siltstones and coals worked in several of the Asturian Coalfields; in detail the stratigraphy is complex.

In the León-West Asturian Zone Precambrian occurs in the Narcea Antiform on its E. boundary, in a large antiform within the zone and in the narrow Ollo de Sapo "porphyroid" band on its W. margin. Above a sub-Cambrian unconformity the rest of the zone consists of a Lower Palaeozoic succession, including turbidites and pelites, of a deeper-water facies than that of the Cantabrian Zone. The León-West Asturia Zone has undergone low-grade Hercynian metamorphism, and post-tectonic granites have produced contact-aureoles. Movements of uplift and subsidence mainly occurred from the late Precambrian to the Upper Devonian. Wagner and Martinez-Garcia (1974) have stated, however, that an important pre-Wenlock break occurs.

Hercynian (sensu lato) movements began with a Bretonic phase, as Upper Famennian sandstones rest discordantly on lower formations in the Cantabrian Chain. The break is more marked in the N. and Central Asturias than in the S. part of the Chain, referred to by some authors as the Leonides. The latter are held to terminate to the N. at the "León Line", the significance of which has, however, been disputed.

This is the type-locality for the Asturian movements and the main folding is dated as early Stephanian. However, earlier phases, from the beginning of the Westphalian onwards, have been recognised, although there is controversy regarding their relative significance (de Sitter, in: Coe, 1962; Wagner, 1963; Wagner and Martinez-Garcia, 1974). The Westphalian clastics in the S., referred to above, probably reflect these earlier disturbances; further N. the Asturian belt was sinking, leading to the deposition of thick manganese-shales.

The Cantabrian Arc is a most striking feature of the León-West Asturian Zone.

The folds swing from E.-W. in the N., where they can be well seen on the coast near Llanes, through N.-S. in the W. to W.-E. in the S., N. of León. The sharp swing in the strike of the U. Westphalian Coal Measures is clearly seen in the mining valley of the Lena, S. of Oviedo. As the Stephanian basins also curve with the Arc, much, at any rate, of its development was a very late Hercynian event. Wagner and Martinez-Garcia (1974) suggest that the formation of the Arc was controlled by a westwardly projecting foreland massif.

In the Bernesga Valley, Asturian compression formed five thrusts towards the N., i.e. towards the centre of the Arc. Each nappe contains a succession from the Lancara formation (Cambrian) to the Visean (Namurian). Julivert (1974) holds that the Lancara formation formed a décollement surface. Cross-folds on N.W. axes also occur.

Further W., in the Luna Valley, the structure becomes simpler. De Sitter and Van den Bosch (1968) have described the thrusting of Cambrian over Lower Carboniferous, as can be seen in the craggy hills on the S. margin of the Chain, N. of León. The later Stephanian molasse sediments, with their coals, were themselves folded and faulted by a late Hercynian phase, dated as Saalian. At least one thrust was developed, that of San Mateo, which brought Devonian limestone over Stephanian.

Hercynian mineralisation resulted in the deposition, mainly along faults, of ores of copper, nickel, cobalt, iron, mercury and lead; nearly all the mines are now abandoned.

In Middle Tertiary times the Cantabrian Chain was uplifted as reversed faulting and folding of the Cretaceous developed on its S. margin. A major Tertiary dextral transcurrent fault, cutting S.E. through the Chain from the coast N. of Oviedo, may be connected with the Biscay opening.

5.14 The Central Iberian, the Ossa-Morena and the South Portuguese Zones (Figs. 5.8 and 5.9)

The Central Iberian Zone is the axial zone of the Iberian Hercynides (Aubouin, 1965, 213; Bard et al., 1973, 50). It forms the high (500 - 1000 m), often remarkably flat, meseta of much of Spain and Portugal, diversified by several mountain ranges; the highest of these is the Sierra Gredos, N.W. of Madrid, where the Pico de Almanzor rises to 2592 m.

Good exposures occur on the higher ground, but on the meseta the Hercynian rocks form scattered, deeply weathered outcrops and are often covered by Tertiary strata with relatively thin intervening Mesozoic in places. Excellent sections can, however, be seen on the N.W. Spanish coast and on the Portuguese coast N. of Oporto.

Older Precambrian rocks, which have undergone Precambrian metamorphism, appear both as "basement" and as remnants of a thrust plate (see below). A thick, clastic flyschoid sequence ranges from Precambrian to late Cambrian. This is overlain unconformably by Ordovician to Middle Devonian shelf deposits. These are followed by transgressive marine Lower Carboniferous, above which comes an interrupted succession of Upper Carboniferous sediments. Much of the zone consists of large granite massifs, mostly post-tectonic, although some may be late syntectonic as suggested by their structures. Some pre-Hercynian intrusions also occur.

Fig. 5.9 A. Migmatite. Near Garabit Viaduct, Massif Central, France.
B. Pelite with andalusite crystals, Central Iberian Zone. North of Viana, Portugal.

In the older Precambrian rocks, Precambrian structural and metamorphic events are recorded, over-printed by Hercynian structures and metamorphics. Tectonic movements occurred at the end of the Cambrian, as shown by the sub-Ordovician unconformity, but it is doubtful if this amounted to a Caledonian orogenic phase.

At least two Hercynian phases can be made out, one pre-Visean, i.e. Bretonic, the other pre-Stephanian, i.e. Asturian; the Stephanian, in fact, is discordant on all the older formations. Accompanying metamorphism has also been shown to be polyphase and in the higher grades is often of andalusite/ kyanite/sillimanite type (Bard, 1973, 60).

It is often difficult to separate the fold-phases, but the pre-Visean folds can be shown to have vertical or, rarely, westerly-recumbent axial planes and the Asturian folds, with longer wavelength, generally vertical planes. The strike swings from S. in the N. to S.E. in the S.

In the N.W. of the Zone, Precambrian rocks occupy a number of synforms, e.g. those of Cabo Ortegal, Ordenes, Bragança and Morais. These rocks are of catazonal grade and include augen-gneiss, polymetamorphic amphibolite, granulite facies rocks and eclogite. They overlie Silurian metasediments of green-schist facies. They are interpreted (Ries and Shackleton, 1971) as parts of a plate of Precambrian thrust up eastwards from the deep crust during the Hercynian orogeny. The Zone is cut by several major N.E. and N.W. faults cutting all the Hercynian structures; similar transcurrent movement is claimed for some of the N.E. faults.

In early Tertiary times slight buckling formed the depressions in which the Tertiary sediments were deposited; the considerable thicknesses of these sediments, which go down to 1500 m below sea-level in the Douro Basin, show that depression continued in the early Tertiary. The Tertiary sediments were slightly folded in places, probably by mid-Tertiary compression, and renewed movement occurred along some of the older faults.

The Ossa-Morena and South Portuguese Zones form meseta and sierra, rising to about 1000 m, and underlie the Portuguese coastal plain. The South Portuguese Zone outcrops on the rocky Algarve coast.

The Hercynian structures very dominantly strike S.E. with overturning towards the S.W. and the two zones, although not precisely similar, form a "mirror-image" of the two northerly zones (5.13). The contact with the Central Iberian Zone, S.E. from Oporto to Córdoba, is a fault, which may be transcurrent.

Precambrian rocks, which have undergone metamorphism up to the kyanite-sillimanite grade, are abundant in the Ossa-Morena Zone but absent in the South Portuguese Zone. Otherwise, both contain a Cambrian to Upper Carboniferous succession with local breaks. Post-tectonic granites are abundant in the Ossa-Morena Zone but rare in the other.

In the Ossa-Morena Zone there were important structural and metamorphic events, probably Cadomian, before the deposition of the first Cambrian beds (Bard et al., 1973, 61). The pelites were raised to the kyanite/sillimanite grade and later underwent partial or total Hercynian retrograde metamorphism. Much of the Hercynian folding was pre-Visean (Bretonic) and the latest phase seems to have been pre-Middle Westphalian, with regional metamorphism reaching its peak between the two.

In the South Portuguese Zone the main Hercynian folding and metamorphism was later, probably Asturian, and was accompanied by greenschist metamorphism - a difference from the Cantabrian Zone. Late movement formed nappes with overthrusting towards the S.W. The Iberian Pyrite Belt of ore deposits, in Lower Carboniferous, extends through the Zone.

The western part of both Zones is overlapped by a Mesozoic/Tertiary succession. This reaches a depth of over 4000 m, showing that large-scale subsidence took place. These younger strata are cut by a number of faults, some of which also cut the Palaeozoic rocks. Several have been traced across the Continental Platform to the Continental Slope.

REFERENCES

Aubouin, J. 1965. Geosynclines. Elsevier, Amsterdam.
Bard, J.P. 1973. Les ceintures metamorphiques de la Meseta iberique. Réun. Ann. des Sci. de la Terre. Programmes Résumés. Paris.
Bard, J.P. et al. 1973. Données préliminaires sur l'extension et les caracteres de la ceinture metamorphique de Badajoz-Córdoba (Sierra Morena occidentale, Espagne). Réun. Ann. des Sci. de la Terre. Programmes Résumés. Paris.
Bishop, A.C. et al. 1975. Precambrian rocks within the Hercynides. In: The Precambrian. Spec. Rep. No. 6, Geol. Soc.
Bott, M.H.P. et al. 1958. The geological interpretation of gravity magnetic surveys in Devon and Cornwall. Abs. Roy. Geol. Soc., Cornwall.
Chenevoy, M. 1974. Le Massif Central. Géologie de la France. (ed. J. Debelmas). Paris.
Dearman, W.R. 1959. Easter Field Meeting in North-West Dartmoor. P.G.A., 70, 338-341.
Dearman, W.R. 1963. Wrench-faulting in Cornwall and South Devon. P.G.A., 83, 265-287.
Dearman, W.R. 1969. An outline of the structural geology of Cornwall. Proc. Geol. Soc. London, No. 1654, 33-39.
Dearman, W.R. 1970. Some aspects of the tectonic evolution of South-West England. P.G.A., 81, 483-491.
Dearman, W.R. et al. 1970. The north coast of Cornwall from Bude to Tintagel. Exc. Guide No. 10. Geol. Assoc.
De Sitter, L.V. In: Coe, 1962. The Hercynian Orogenes in Northern Spain. University Press, Manchester.
De Sitter, L.V. & van den Bosch, W.J. 1968. The Structure of the S.W. part of the Cantabrian Mountains. Leid. geol. Meded., 43, 213-216.
Dodson, M.H. & Rex, D.C. 1971. Potassium-argon ages of slates and phyllites from South-West England. Q.J.G.S., 126, 465-499.
Fourmarier, P. 1950. Compte Rendue de la Session Extraordinaire de la Societe Geologique de Belgique. Ann. Soc. Geol. de Belg., 73, 151-218.
Gèze, B. 1960. L'orogénèse caledonniene dans la Montagne Noire. Geol. Congr., 1960, 19, 120-125.
Gill, W.D. 1962. In: Some Aspects of the Variscan Fold-Belt, (ed. K. Coe). University Press, Manchester.
Graindor, M.J. 1961. Géologie du nordouest du Cotentin. Bull. Serv. Carte Geol. France, 272.
Hall, A. 1974. West Cornwall. Exc. Guide No. 19. Geol. Assoc.

Hobson, D.M. & Sanderson, D.J. 1975. Major early folds at the southern margin of the Culm synclinorium. J.G.S., 131, 337-352.
Holwill, F.J.W. et al. 1969. Summer (1966) Field Meeting in Devon and Cornwall. P.G.A., 80, 43-62.
Julivert, M. 1971. Décollement tectonics in the Hercynian Cordillera of North-West Spain. Amer. J. Sci., 270, 1-29.
Klein, C. 1963. A propos de la carte géologique de Cherbourg au 50000e. Compt. Rend. Soc. Geol. France, 1963: 309-311.
Lambert, J.L.M. 1964. The unstratified sedimentary rocks of the Meneage area, Cornwall. Circular 118, Proc. Geol. Soc. London, 3.
Michot, J. 1955. Génèse du chloritoide en Milieu statique. Ann. Soc. Geol. de Belgique, 78, 3-54.
Miller, J.A. & Green, D.H. 1961. Preliminary age-determinations in the Lizard area. Nat., 191, 151.
Owen, T.R. 1971. The relationship of Carboniferous sedimentation to structure in South Wales. Comp. Rend. de Congres Intern. Strat. Geol. Carbonif. Sheffield 1967, III, 1305-1316.
Owen, T.R. In: Nairn & Stehli (eds.). 1974. The Geology of the Western Approaches. The Ocean Basins and Margins. Vol. 2. The North Atlantic. Plenum Press, New York.
Pelletier, H. 1969. Notes Historiques sur les Seismes en Auvergne. Rev. des Sci. Nat. d'Auvergne, 35, 23-32.
Peterlongo, J.M. 1972. Massif Central, Limousin, Auvergne, Velay. Guid. Géol. Rég. Masson et Cie, Paris.
Philcox, M.E. 1963. Compartment deformation near Buttevant, Co. Cork, Ireland, and its relation to the Variscan thrust front. Sci. Proc. Roy. Dublin Soc., A2, 1-11.
Ries, A.C. & Shackleton, R.M. 1971. Catazonal Complexes of North-West Spain and North Portugal. Remnants of a Hercynian Thrust Plate. Nat. Phys. Sci., 234, 65-68 and 79.
Rutten, M.G. 1969. The Geology of Western Europe. Elsevier, Amsterdam.
Roques, M. et al. 1954. Summer Field Meeting in the Massif Central. P.G.A., 65, 278-312.
Tilmann, N. et al.1938. Summer Field Meeting 1937: The Rhenish Schiefergebirge. P.G.A., 49, 225-260.
Wagner, R.H. 1963. A general account of the Palaeozoic Rocks between the Rivers Porma and Bernesga (León, N.W. Spain). Boletin del Instituto Geologico y Minera de Espana, 74, 1-159.
Wagner, R.H. & Martinez-Garcia, E. 1974. The Relation between Geosynclinal Folding Phases and Foreland Movements in North-West Spain. Studia Geologica, 7, 131-158.
Wallace, P. 1972a. The Geology of the Palaeozoic Rocks of the South-Western Part of the Cantabrian Cordillera, North Spain. P.G.A., 83, 57-73.
Wallace, P. 1972b. Summer Field Meeting in the Cantabrian Cordillera, North Spain. P.G.A., 83, 75-94.
Wallace, P. & Laurentiaux, D. 1973. Summer Field Meeting in the Ardennes and Vosges. P.G.A., 84, 181-206.
Walsh, P.T. 1968. The Old Red Sandstone West of Killarney, Co. Kerry, Ireland. P.R.I.A., 66B, 9-26.
Waterlot, G. et al. 1973. Ardenne. Guides Géol. Rég., Paris.

Chapter 6
HERCYNIAN SEDIMENTARY BASINS, BLOCKS AND COALFIELDS

6.1 Introduction

A Hercynian front crossing the S. of the British Isles (5.11) forms the northern limit of strata overfolded, overthrust and metamorphosed during the Hercynian orogeny and of post-tectonic Hercynian granites. The front does not, however, mark the northern limit of Hercynian tectonism, for almost all pre-Permian rocks of the British Isles show, to some extent at least, Hercynian structures. In pre-Devonian blocks, rendered resistant by Caledonian or older events, these structures are mainly faults, often transcurrent. In the Upper Palaeozoic terrains, however, N. of the front, Hercynian folds and faults are present everywhere although overfolds and overthrusts are rare. The terrains affected by this "external" Hercynian tectonism include virtually all the British coalfields. The following Hercynian blocks, basins, etc., will be considered: the Midland Valley of Scotland (6.2), the North-East of England (6.3), Central England (6.4), the South Wales and Pembrokeshire coalfields (6.5), the Bristol coalfield (6.6) and Central Ireland (6.7).

6.2 The Midland Valley of Scotland (Figs. 6.1 and 6.2)

The Midland Valley is a complex graben, bordered to the N.W. by the Highland Boundary Fracture-zone (3.3) and to the S.E. by the Southern Uplands Fault (4.4). Although the greater part of the region lies below the 500 ft (152 m) level, it is not a valley in the ordinary sense of the term, but consists of several low-lying areas diversified by hills mostly formed of igneous rocks. The highest point is Ben Clough, 2363 ft (720 m) in the Ochil Hills. The hills and the E. and W. coasts provide good exposures. The low ground is largely mantled in glacial drift, but there are numerous working and disused quarries. Excursions have been described by Bassett (1958), Macgregor (1968) and Craig and Duff (1976).

The geological succession is as follows:-

```
                Triassic
                Permian

                Coal Measures               )
                Millstone Grit              )
                Scottish Carboniferous      )   Carboniferous
                Limestone Series            )
                Calciferous Sandstone Series)

                Upper Old Red Sandstone     )
                Lower Old Red Sandstone     )   Devonian

                        Silurian
                        Ordovician
```

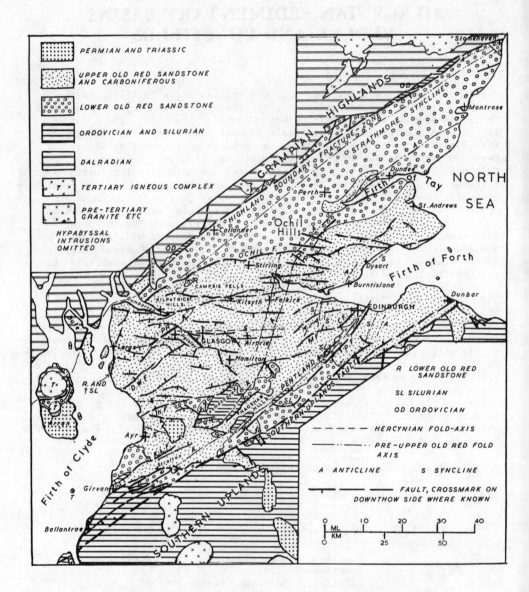

Fig. 6.1 STRUCTURAL MAP OF THE MIDLAND VALLEY OF SCOTLAND
C.F. = Campsie Fault; D.W.F. = Dusk Water Fault;
In.F. = Inchgotrick Fault; K.L.F. = Kerse Loch Fault;
MF/CF = Murieston/Colinton Fault; P.F. = Pentland Fault;
P.R. = Paisley Ruck

The Arenig consists of an ophiolite association of spilite lavas, pyroclastics, black shales and cherts. The Upper Ordovician and the Silurian are made up of highly fossiliferous sedimentary rocks of shelf facies.

In both the Lower and Upper Old Red Sandstone deltaic, fluviatile and lacustrine sediments predominate. Thick andesite and basalt volcanics are present in the lower series.

The Scottish Lower Carboniferous contrasts strongly with that of the rest of Britain, and there are major facies variations within the Midland Valley itself related to older structures. The Calciferous Sandstone Series starts with a lagoonal succession of alternating shales and ferro-dolomites, followed in the W. by thick basalt lavas - the Clyde Plateau Lavas - and in the E. by thinner lavas of both basaltic and acid character. High up in the series the workable Oil-Shale Groups occur in the E., whereas in the W. there are varied sediments including thin marine limestones. In the Scottish Carboniferous Limestone the thick carbonate rocks of much of the Western European Dinantian are absent; only thin limestones occur, along with shales, sandstones and a deltaic coal-bearing succession which is extensively worked. The Coal Measures do not range above the Westphalian; valuable seams occur in the lower part but the upper consists of red sandstones and shales. The Permian and Triassic are of New Red Sandstone facies with basic lavas in the Permian. The region contains intrusives of Arenig, Lower Old Red Sandstone, Lower Carboniferous, Permian and Tertiary age.

Tectonic Evolution

Structural events were :-

    Caledonian   (Pre-Devonian
                       (Mid-Devonian

    Hercynian   (Pre-Namurian (Sudetic) - mainly uplift
                      (Post-Westphalian (Asturian)

                       Post-Permian - probably Tertiary - faulting

There is proof of mid-Ordovician folding in the Girvan-Ballantrae Lower Palaeozoic outcrop in the S.W. of the Midland Valley, where there is an unconformity between the Ballantrae Igneous Series and conglomerate at the base of the Baer Series, generally accepted as Caradoc. Folding took place predominantly along N.E. axes.

The main folding of the Southern Uplands (4.4) was late-Silurian; there is also clear evidence of late-Silurian/pre-Devonian folding within the Midland Valley in the Girvan and Pentland Hills areas. On the other hand, in the Lesmahagow district there is no break between the Silurian and Devonian and it may be concluded that these overlie a block which resisted this compression.

Following the work of Lapworth (1882) and Peach and Horne (1899), folding in the Girvan district, affecting strata up to the Wenlock, was regarded as largely isoclinal on N.E. axes. Williams (1959, 1962), however, put forward the view that isoclinal folding is unimportant and that the succession is much thicker than previously believed. Five phases of deformation are

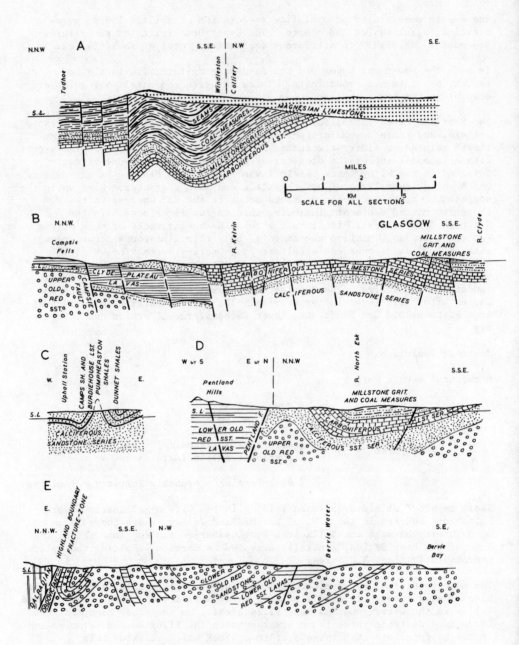

Fig. 6.2 SECTIONS OF NORTH-EAST ENGLAND AND OF THE
MIDLAND VALLEY OF SCOTLAND
  A. Durham Coalfield
  B. Campsie Fells to Glasgow.
  C. Pumpherston District, Lothians Oil-Shale Field
  D. East side of Pentland Hills
  E. North-East of Midland Valley

recognized.

By Lower Devonian times the Midland Valley must have been a Caledonian molasse basin, formed by a combination of faulting and downwarping. Lower Old conglomerates on its N.W. side contain Highland boulders and those on the S.E. margin Southern Uplands boulders.

The region provides the most striking expression in the British Isles of mid-Devonian deformation, for throughout the region and its margins powerful folding and faulting took place on north-easterly axes, and the Upper Old Red Sandstone rests with strong unconformity on the Lower or on older strata. The Strathmore Syncline, affecting the great thickness of Lower Old Red strata outcropping in the N.E., has an amplitude of at least 15,000 ft and is traceable from the E. coast for 110 miles to Loch Lomond.

The Pentland Hills Anticline extends south-westwards from Edinburgh for at least 25 miles and on the continuation of the same axis to the S.W., for a total distance of 80 miles, there are the Carmichael, Hagshaw Hills, and Straiton upfolds.

The main displacements along the Highland Boundary Fracture-Zone (3.3) and the Southern Uplands Fault (4.4) took place in mid-Devonian times. By the beginning of the Upper Devonian the Midland Valley was thus a true graben bounded by faults.

The Millstone Grit rests in places on strata as low as the Calciferous Sandstone Series. The angular unconformity is, however, non-existent or slight, showing that the pre-Namurian movements were mainly of uplift.

The most important structures of the Midland Valley are post-Westphalian and were formed by phases of maximum pressure and relief of pressure acting in a N.-S. direction. Nevertheless, the Caledonian borders and the buried Caledonian structures continued to exert a strong influence. Consequently there are two main sets of Hercynian structures, those with a "normal" E.-W. trend and those with a N.-E. to N.N.E. or caledonoid trend. The interlocking nature of the two sets is epitomized by the complex structural depression of the Central Coalfield which has both an E.-W. axis and a N.N.E. axis. East of the latter, in fact, caledonoid structures predominate in the Carboniferous rocks; to the W., on the other hand, E.-W. structures are more common. The Carboniferous strata of the west central Midland Valley, including the Glasgow district, form a broad syncline pitching E. The syncline is greatly complicated by many E.-W. folds and faults; most of the faults step down to the S., N. of the Clyde and to the N., S. of the river. The Campsie Fault has a downthrow to the S. which reaches a maximum of nearly 3000 ft (915 m). The Coal Measures of the Upper Carboniferous extend N.N.E. to Clackmannanshire where they are cut off by the Ochil Fault, bringing up Lower Devonian lavas, a displacement of at least 10,000 ft (3050 m).

From oil-shale workings W. of the Pentland Hills, numerous sharp folds have been proved with axes striking from N.N.E. to N. by E.

The Pentland Hills Anticline is primarily a mid-Devonian structure but it was reactivated by Hercynian compression, as the Upper Old Red Sandstone and Lower Carboniferous are folded round its northern end. The Pentland Fault, with displacement of several thousand feet, can be observed to be reversed and, in fact, has been shown to have a reversed hade of about 22° in a bore (Lees and Taitt, 1946, 275). The fault separates the anticline from a zone of marked N.N.E. folding in the Carboniferous strata to the S.E.

The Midland Valley provides a good example of "folding within a frame" and the production of north-easterly folds near its margins by Hercynian N.-S. compression can be ascribed to the influences of the already upfaulted blocks flanking the graben. The N.E. to N. by E. folds of the central areas, however, require further discussion. It seems probable that the caledonoid structures exercised basement control, the anticlines not only being accentuated but being forced closer together and thus producing from the primary N.-S. Hercynian compression a resolved compression which extended upwards through overlying Carboniferous sediments. In the case of the folds E. of Edinburgh the controlling structures were probably the Southern Uplands and the Pentland Hills Anticline. For those W. of Edinburgh the controls may have been the Pentland Hills Anticline and a Caledonian anticline now completely buried by the sediments of the eastern part of the Central Coalfield. The existence of such a structure is supported by isopachyte studies by Kennedy (1958, 121-4) and by the presence of igneous rocks of Lower Old Red Sandstone type at from 3990 to 4267 ft (1220-1300 m) in a hole at Salsburgh, E. of Airdrie (Falcon and Kent, 1960).

Faults on an E.-W. trend cut the highest Carboniferous rocks; some of these movements were earlier, some later, than the intrusion of conspicuous E.-W. quartz-dolerite dykes, associated with thick sills of the same composition.

A number of powerful N.E. faults cut the interior of the Midland Valley and, while these can be interpreted as normal (or in the case of the Pentland, reversed) faults, they may well be, as pointed out by E. M. Anderson (1951), transcurrent faults or at any rate have an important strike-slip component. Under N.-S. compression the relatively rigid basement would tend to yield by shearing rather than by folding and because of the Caledonoid grain there would be a preferential development of N.E. shears.

In Arran folding and faulting affect Triassic sediments and Tertiary igneous rocks. Some of this is due to Tertiary igneous events (9.2) but structures in the southern part of the island, including a number of N.N.W. faults, are more likely to be of normal tectonic origin.

Some N.W. faults, particularly in the W. of the Midland Valley, have been ascribed to the Tertiary, although many of these could be older or be due to reactivation of pre-existing fractures. Tertiary dyke intrusion too is evidence of tensional fissuring.

6.3 North-East England (Figs. 6.2 and 6.3)

North-east England is the region bounded to the W. by the Pennine and Dent faults, to the S. by the North Craven Fault and to the E. by the base of the Permian. It contains the northern part of the Pennines rising to 2930 ft (893 m) at Cross Fell and the high hills on the Scottish Border including the Cheviot, 2676 ft (815 m). Glacial deposits are heavy on much of the region, but there are good coastal and river sections. Excursions are described by Tomkeieff (1965) and by Johnson (1973).

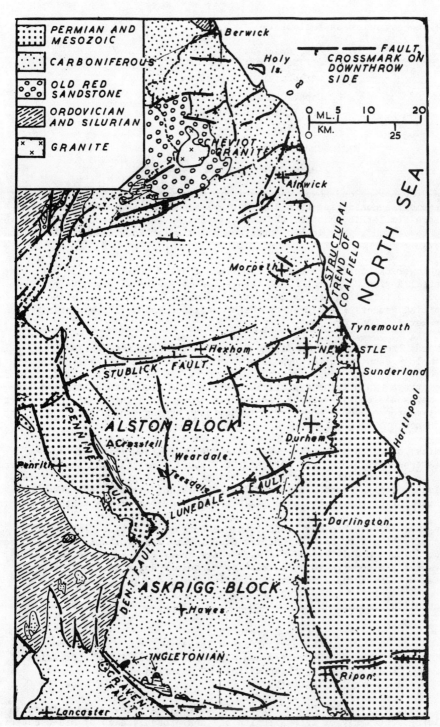

Fig. 6.3 STRUCTURAL MAP OF NORTH-EAST ENGLAND

The succession is as follows :-

Coal Measures)	
Millstone Grit)	Carboniferous
Carboniferous Limestone)	
Upper Old Red Sandstone)	Devonian
Lower Old Red Sandstone)	
Silurian	
Ordovician	
Ingletonian - ? Precambrian	

The Precambrian and Lower Palaeozoic form very limited outcrops, the Precambrian consisting of slates and grits and the Lower Palaeozoic of strata similar to those of the same age in the Lake District (4.3). Both the Lower and Upper Old Red Sandstone are similar to the Scottish facies (6.2).

The Lower Carboniferous succession links that of Scotland (6.2) with that of Central England. In the N. of the region it starts with the Cementstone Group - alternating shales and ferro-dolomites (with basalts along the Scottish Border) - followed by thick sandstone and by the workable Scremerston Coal Group.

Towards the S.W., that is at the N. end of the Pennines, marked changes in lithology occur with the appearance of more abundant marine limestones. The Scremerston Coal Group, as such, disappears and at about the same horizon the strata exhibit the typical Yoredale rhythm with a repetition of limestone, shale and sandstones, frequently topped by a thin coal seam. Further S. limestones become more and more dominant and spread to the base of the sequence.

Shales, siltstones and fine-grained sandstones, identified on maps as Millstone Grit, and lying between the highest limestone and the lowest coal of the Coal Measures, form large outcrops in the region but until more palaeontological evidence is available it is not certain that these are equivalent to the established Millstone Grit Series of the southern Pennines.

The Coal Measures in Northumberland and Durham are divided into a Lower Group, a Middle or Main Productive Group, with many workable seams, and an Upper Group, mostly sandstones with a few coals. Rocks up to the Coal Measures are penetrated by intrusions of quartz-dolerite, including the Great Whin Sill and a number of dykes.

Tectonic Evolution

> ? Late Precambrian
> Caledonian (sensu lato)
> Dinantian and Namurian - uplift and depression
> Post-Westphalian - Asturian
> Tertiary

The small outcrops of Lower Palaeozoic rocks show evidence of Caledonian folding, certainly of late Silurian age and probably also of Mid-Ordovician age. North-westerly isoclinal folds in the Ingletonian rocks, contrasting with the Caledonian structures, are evidence of a late Precambrian structural event. According to some radiometric dates, however, the Ingletonian may not be earlier than Cambrian (O'Nions et al., 1973).

The Lower Old Red Sandstone volcanic pile around the Cheviot is invaded by the large Cheviot granite, the intrusion of which may have been accompanied by uplift. Circumferential tension is shown by the presence of a surrounding dyke-swarm on a roughly radial pattern. Mid-Devonian movement, as in the Midland Valley of Scotland (6.2), is evident from the unconformable overlap of the Upper Old Red Sandstone on the W. side.

Further S. the Alston and Askrigg Blocks, perhaps already brought into existence by the late-Silurian folding, were further shaped by Mid-Devonian movements and by the intrusion of granites below the present erosion surface. The Weardale Granite, in the Alston Block, was proved by the Rookhope bore (Dunham et al., 1965) to lie unconformably beneath Visean at 1281 ft (390 m) and the Wensleydale Granite, in the Askrigg Block, was shown by the Raydale bore (Dunham, 1974) to occur at 1625 ft (495 m) beneath Visean. Both have ages around 400 m.y. Dinantian and Namurian sedimentation provide evidence of repeated distortion of the pre-Carboniferous floor. Thus the Lower Carboniferous sediments, up to 8000 ft (2420 m) N. of Newcastle, thin to well under 1700 ft (520 m) over the Alston and Askrigg Blocks and thicken to 6000 ft (1828 m) beyond the Craven Faults S. of the Askrigg Block.

Movements of the Alston and Askrigg Blocks continued into the Namurian because over these structures the Millstone Grit is thin and interrupted by internal unconformities. There is no evidence, however, that the Alston-Askrigg Block had marked effects on Westphalian sedimentation, nor is there any sign in the stratigraphy of the Coal Measures of an anticlinal Pennine Axis (George, 1963, 46).

The main deformation was probably late-Westphalian and certainly pre-Permian. These movements had less effect on the Carboniferous of the Alston and Askrigg Blocks. The Coalfield Basin has a N. by E. or sub-caledonoid trend with the western limb dipping fairly gently off the Pennine anticlinal axis. The eastern limb rises under the Permian with, in Durham, flanking folds of a smaller order. On the E. side of the Durham Coalfield, in fact, the Coal Measures are cut out by the sub-Permian unconformity and the Magnesian Limestone rests on the Carboniferous Limestone. Near Berwick, Shiells (1964) has described a major E.-facing isocline and has pointed out that in this district folds generated by E.-W. compression are deflected round the Cheviot and Southern Upland masses.

The Coalfield is traversed by numerous faults, nearly all of "normal" type. The major faults range within $25°$ of E. and W.; there is a second set trending N. by W. which is of less significance. Important Hercynian movements also took place along the faults bounding and cutting the Alston-Askrigg Block. The Pennine Fault (4.3) was initiated by Hercynian movement.

The Lower Carboniferous has undergone epigenetic lead - zinc - fluorine barium mineralisation, and the region was at one time an important lead/zinc mining field. Significant mineralisation is concentrated in the relatively thin sediments of the Alston and Askrigg Blocks (Dunham, 1967).

Post-Liassic, probably Tertiary, movement led to the final development of the Pennine arch and the Pennine Fault. A fault at the E. margin of the region, S.W. of Hartlepool, is post-Triassic and almost certainly Tertiary, and it is possible that further faulting in the coalfield took place at the same time.

6.4   Central England (Figs. 6.4 and 6.5)

Although Hercynian events largely shaped Central England, Alpine movements played a greater role than was the case further N.; moreover, older, Precambrian rather than Caledonian, structures exercised a significant influence on the development of both the Hercynian and Alpine structures. The region, as defined for the present purpose, lies S. of the Craven Faults and W. of the base of the Permian as far S. as Nottingham. From there an arbitrary line, running S. then S.W. to near Gloucester, separates the present region from S.E. England (8.8), and another W.N.W. line to beyond Hereford divides the Old Red Sandstone of Central England from that described in the next section (6.5). Northwards, to the sea at the Dee Estuary, Central England borders the Welsh Block, partly along the Church Stretton line (4.2).

Along the Pennine upfold the ground rises to over 2000 ft (610 m) and hard outcrops further S., e.g. the Precambrian of the Malvern Hills, rise to over 1000 ft (305 m), but much of the region, which includes the Midland Plain, lies below 400 ft (122 m). Exposures on this lower ground and along the coast are sparse, owing to the widespread glacial drift, but there are numerous quarries.

The geological succession is as follows :-

>   Liassic
>   Triassic
>   Permian
>   Coal Measures
>   Millstone Grit
>   Carboniferous Limestone
>   Old Red Sandstone
>   Silurian
>   Cambrian
>   Precambrian

The Precambrian appears in a number of small inliers and probably underlies the S. of the region as part of Precambrian kratogen separated by the Church Stretton Zone (4.2) from the Caledonian fold-belt. The latter, sweeping round the Precambrian block, may, however, underlie the N. of the region and continue under eastern England (8.8).

The Precambrian outcrops fall into three groups :-
(a) the plutonic, dominantly intermediate and locally gneissose rocks with associated schists (the last apparently confined to shear zones) of the Malvern Hills, and the schists of Rushton;
(b) the slates, quartzites and pyroclastics of Charnwood; and
(c) the volcanics of the Lickey Hills, Nuneaton, etc.

The age-order is probably from (a) to (c) but this has been disputed. Group (c) is probably equivalent to the Uriconian of the Church Stretton area

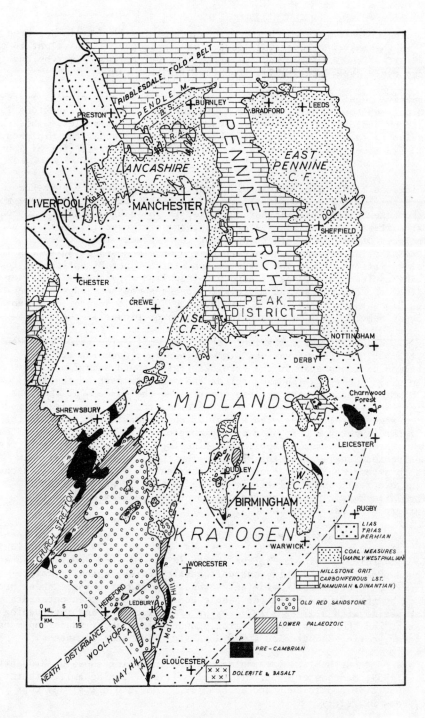

Fig. 6.4 STRUCTURAL MAP OF CENTRAL ENGLAND

(4.2). The Charnian, in which the problematical fossil Charnia has been found, is the most controversial (Ford, 1963). It may be equivalent to the oldest part of the Longmyndian, to part of the Uriconian, or may belong to an older cycle (see Baker, 1971; Dunning, 1975).

The Cambrian consists of quartzites and shales; the Ordovician is represented only by the Caradoc, and, in the Silurian, Llandovery sandstone is followed by Wenlock and Ludlow limestones and calcareous shales. This shelf succession is further evidence for a kratogenic block under the Midlands. The Old Red Sandstone sequence is continental and similar to that of South Wales (6.5).

Carboniferous rocks cover an extensive portion of the surface over the northern half of the region as far S. as the Peak District, but become more isolated beneath the Permo-Triassic blanket in the southern half. In the N. the Carboniferous succession reaches a thickness of over 17,000 ft in places and includes Dinantian strata (down to at least basal Visean levels), a complete Namurian sequence and a virtually complete Westphalian succession. Lateral variations within the basin facies of the Upper Dinantian and Lower Namurian resulted in spreads of sand or of Bowland-type shales. Further N. the effect of contemporaneous movements along the southern edge of the Askrigg Block resulted in only a thinner high Visean sequence, oversteps in the lowest Namurian and perhaps a localization of Dinantian reefs.

To the S., also, lateral variations set in, the main kind being the result of overlap or overstep of various higher Dinantian, Namurian and even Westphalian horizons against the irregular pattern and behaviour of a Mercian or St. George's landmass across the southern part of the area. The basal Namurian becomes more or less absent into the borders of the Peak District, whilst much further S. near Wolverhampton the Dinantian is reduced to a few feet in thickness. The effect of the Dinantian overstep on to older rocks is perhaps best seen on the margins of the Little Wenlock Coalfield in Shropshire, the Carboniferous Limestone eventually coming to rest on the Tremadoc shales near the Wrekin. The thin Dinantian appears in places to extend southwards even beyond the feather edge of the Millstone Grit and eventually by the latitude of the Clee Hills and the south Staffordshire-Warwickshire Coalfields, the transgressive base of the Carboniferous is formed by the Coal Measures.

The Permian and Triassic are difficult to separate and are often collectively referred to as the New Red Sandstone. In fact, the highly variable and locally thick formations of the latter in the Midlands bear evidence of continued unrest. The Permian, as defined within the region, consists dominantly of continental, arid-facies sandstone and breccias; these are followed by the (Triassic) Bunter Pebble Beds. The area then entered a quieter period with a widespread lowering of surrounding and intervening hill areas and the deposition of firstly non-marine Keuper sands and muds and, later, marine Rhaetic and Lower Jurassic clays and carbonates.

There are numerous publications on excursions including those by Broadhurst et al., (1959, Manchester), Garrett et al., (1960, Birmingham), Bathurst et al., (1965, Liverpool), Penn and French (1971, Malvern Hills) and Cope (1976, Peak District).

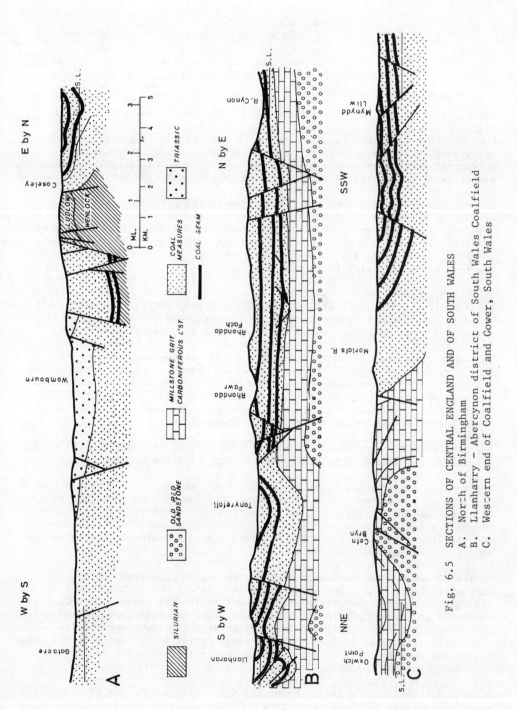

Fig. 6.5 SECTIONS OF CENTRAL ENGLAND AND OF SOUTH WALES
A. North of Birmingham
B. Llanharry – Abercynon district of South Wales Coalfield
C. Western end of Coalfield and Gower, South Wales

Tectonic Evolution

                        Precambrian
                        Caledonian (sensu lato)
                        Pre-Namurian (Sudetic)
                        Post-Westphalian (Asturian)
                        Permian - mainly uplift
                        Alpine

Because of the problems of stratigraphical correlation and dating it is difficult to establish a sequence of Pre-Hercynian events. It is clear, however, that these led to the establishment of three structural trends, malvernoid (N.), charnoid (N.W.) and caledonoid (N.E.). The region is noteworthy for the extent to which these trends, and, in fact, the "basement", generally played a part in later movements right up to the Tertiary.

The Malvern Hills form a narrow ridge, 4½ miles (7 km) long, in a N.-S. direction, forming part of the much longer Malvern Axis which extends S. to well beyond the Severn (6.6). The structural events which affected the Pre-cambrian plutonics and metasediments, and the later Warren House volcanics (probably Uriconian), are themselves Precambrian, but it is not certain that the same movements established the Malvern Axis.

The Charnian rocks form a faulted N.-W. pericline invaded by post-orogenic granophyric diorite (markfeldite) intrusions giving conflicting dates, but probably late Precambrian; to the E. lie the exposed Caledonian Mount Sorrel granodiorite and other concealed intrusions (Le Bas, 1972).

The Lower Palaeozoic Devonian succession E. of the Church Stretton Zone, including the highly fossiliferous Wenlockian limestone of Wenlock Edge, has a N.-E., that is Caledonian, trend. The folding was probably pre-Llandovery (or pre-Caradoc) and mid-Devonian, but was much gentler than in the Welsh Block (4.2). Elsewhere the Caledonian movements were mainly of the block type, further evidence for a Midlands kratogen.

The main structures produced by the combined effects of Hercynian and Alpine movements, influenced by pre-Hercynian trends, will now be summarised.

The Pennine Arch extends S. as far as the Peak District or Derbyshire Dome, bringing up the Carboniferous Limestone. This was an important lead/zinc/fluorspar mining district and, as in the Pennines further N. (6.3), the mineralisation occurs in a relatively thin Lower Carboniferous succession (Dunham, 1967); Lower Ordovician has been proved beneath the Carboniferous at 592 ft (180 m) (Dunham, 1973).

To the E. and W., respectively, of the arch are the East Pennine and Lancashire Coalfields. The important Yorkshire and Nottinghamshire coal mining areas form parts of the East Pennine Field (the largest in Britain), which has an exposed portion of 900 square miles and extends eastwards under the Permian/Mesozoic of Eastern England for at least 2000 square miles. Caledonoid structures include the Rossendale Anticline (bringing up Millstone Grit), the Burnley Syncline and the Knowsley Anticline in the Lancashire Coalfield, and the Don Monocline in the East Pennine Coalfield. Faults in the N.W. or N.E. quadrants have important Tertiary components and are of normal, reversed and "scissor" types (Trotter, 1954). The latter often show rapid diminutions of throw, e.g. from 4000 to 200 ft (1220-61 m) in 3½ miles along the Upholland Fault in the Lancashire Coalfield.

The North and South Staffordshire, Warwickshire, South Derbyshire and Leicestershire Coalfields all occur as separate, large outcrops showing through the New Red Sandstone cover. Some form basins roughly coinciding with the inlier margins; in other cases, e.g. S. of the North Staffordshire Coalfield (Cope, 1954) and between the South Staffordshire and Warwickshire Coalfields, there are considerable concealed coalfields. In South Staffordshire the Coal Measures mostly rest on a Silurian basement which appears in the cores of the sharp Dudley N.N.W. periclines and in smaller domes underground. Malvernian (N.), charnoid (N.N.W. to W.N.W.) and caledonoid (N.N.E. to N.E.) trends are all represented by both folding and faulting.

The Birmingham Fault runs S.S.W. through the New Red Sandstone to the S. margin of the South Staffordshire Field. The N.W. Thringstone Fault, forming the N.E. margin of the Leicestershire Field, brings Cambrian over Coal Measures in depth, but later movements affecting Triassic were in the opposite direction. Large intrusions of olivine-basalt, probably Permian, occur in the Coal Measures near Rowley Regis.

The Malvern Hills are formed from an upthrust strip of older Precambrian plutonic rocks and younger Precambrian volcanics. On the W. side the Precambrian is in reversed fault contact with Silurian folded on N.-S. axes; on the E. it is normally faulted against Triassic. It is generally agreed that the structure is essentially due to Hercynian movements, influenced by basement fractures (Brooks, 1970). Butcher (1962) has suggested that the faulting was preceded by the formation of a monoclinal, Precambrian cored fold with its steep limb to the W. The E. fault has probably a considerable Alpine component, showing that basement influence persisted until this time.

On the southward continuation of the Malvern Axis the May Hill Anticline has Precambrian psammites (similar to those of the Western Longmynd, 4.2) in its core, flanked by Silurian. The E. limb is partly cut off by a fault bringing down Triassic.

To the N.W., the N.W. orientated, much faulted Woolhope Anticline contains a Silurian succession ranging down to the Llandovery. Squirrell and Tucker (1960) have attributed the structures to an E.-W. compressive phase followed by a N.-S. phase, both Hercynian.

The malvernoid, charnoid and caledonoid trends, which are so common in Central England, are anomalous in a N.-S. Hercynian stress-system. They can be attributed to basement control, including folding within basement frames (Owen, 1958). Another influence may have been movement along old, deep basement fractures. For example, the Neath Disturbance (6.5) passes through the hollow (containing a N.E. basic dyke) between the Woolhope Anticline and another small inlier to the N. and continues N.E., probably as far as the Malvern Axis.

6.5 South Wales and Pembrokeshire Coalfields and their Borders (Figs. 6.5 and 6.6)

The South Wales coalfield is the second largest in the British Isles. It is topographically rugged for the hard Pennant sandstone in the Coal Measures forms hills rising to nearly 2000 ft (610 m), deeply dissected by the glaciated South Wales valleys. The surrounding Carboniferous Limestone outcrops in picturesque scarps, and to the N. the Old Red Sandstone reaches

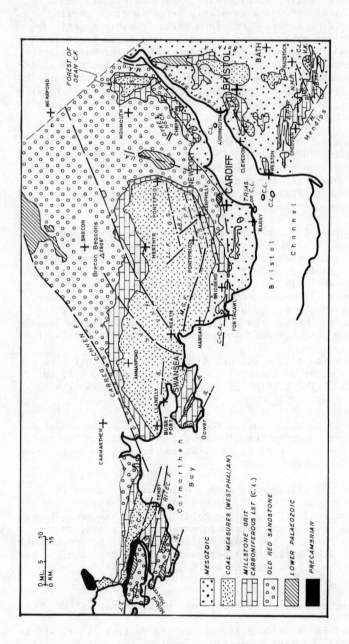

Fig. 6.6 STRUCTURAL MAP OF SOUTH WALES, PEMBROKESHIRE, FOREST OF DEAN AND BRISTOL COALFIELDS.
AB.F. = Aber Fault; B.H.P. = Beacon Hill Pericline; B.P. = Blackdown Pericline; C.-C.A. = Cardiff - Cowbridge Anticline; C.L. = Carboniferous Limestone; C.S. = Caerphilly Syncline; G.S. = Gelligaer Syncline; J.T. = Johnston Thrust; LL.F. = Llanwonno Fault; M.A. = Malvern Axis; M.G.F. = Moel Gilau Fault; N.H.P. = North Hill Pericline; P.A. = Pontypridd Anticline; P.C.F. = Pembrokeshire Coalfield; P.H.P. = Pen Hill Pericline; S.V. = Swansea Valley Fault; TH. = thrust; V.K. = Vobster Klippe; V.N.F. = Vale of Neath Fault.

2906 ft (884 m) in the Brecon Beacons. The Pembrokeshire coalfield, not now worked and separated by Carmarthen Bay, is structurally the continuation of the South Wales field.

An arbitrary line through the Old Red Sandstone, extending N.W. through Hereford, separates the present region from Central England (6.4). The N.W. boundary follows the Church Stretton - Careg Cennen line along the edge of the Lower Palaeozoic Welsh Block (4.2) to Pembrokeshire. Here the S.W. caledonoid trends are affected by the E.-W. Hercynian structures, and the Coal Measures overlap N. onto the Welsh Block.

The South Wales valleys are often deeply floored with drift, as are the coastal platforms, occurring at several levels. Spectacular shore sections occur along the Bristol Channel forming the S. margin of the region.

The succession is as follows :-

>   Lias
>   Triassic
>   Coal Measures
>   Millstone Grit
>   Carboniferous Limestone
>   Old Red Sandstone
>   Silurian
>   Ordovician
>   Precambrian

In the E., Silurian strata occur in the Rumney and Usk inliers where the Wenlock (and possibly the Llandovery in the Rumney inlier) and Ludlow Series consist of shales, siltstones, sandstones (mostly calcareous) and limestones of shelf facies with a rich shaley fauna. In the W., in South Pembrokeshire, Wenlock or Ludlow rests on Llanvirn dark shales. Off the coast the thick, dominantly basaltic volcanics of Skomer Island and the nearby mainland, formerly thought to be Lower Ordovician, are now regarded as Lower Silurian (Ziegler et al., 1969). In Pembrokeshire Precambrian acid volcanics (the Benton Series) are intruded by Precambrian granites and diorites. The Benton Series can be correlated with the Pebidean of St. Davids, discussed in (4.2).

The Lower Old Red Sandstone (Downtonian and Dittonian) around the main coalfield comprise a thick (over 6000 ft, 1830 m) fluviatile succession, consisting of red marls (slightly calcareous siltstones), followed by sandstones and calcareous siltstones with flaggy sandstones (Brownstones). These are directly overlain by the Upper Old Red Sandstone, consisting of the Plateau Beds, with conglomerates, and the Grey Grits; no Middle Old Red has been identified in the region. In the Brecon Beacons the unconformity is not easily seen; the base of the Upper division is marked by a quartz granule bed seen in three boreholes (Taylor and Thomas, 1975). Further E., in the Chepstow area (6.6), there is a clear unconformity and further W., in the Carmarthenshire Fans, there is a slight discordance of dip between the Lower and Upper Series.

Further W. still, the Old Red Sandstone may be over 16,400 ft (5000 m) thick on the N. side of Milford Haven. South of this ria-inlet, which follows the Ritec Fault, a fracture active in Devonian times, the Lower Old Red Sandstone includes wedges of rudites derived from the S. and containing pebbles of Lower Palaeozoic rocks with fossils of North French affinities. The Upper Devonian here consists of a marine sequence, the Shrinkle Sandstone.

Elsewhere the fluviatile or deltaic conditions of the Upper Old Red Sandstone gave place by fairly rapid subsidence, unaccompanied by folding, to the marine conditions of an early Carboniferous sea, flanked to the N., in what is now central Wales, by the ancient massif of St. George's Land. The transitional, although fairly abrupt, change at the junction between the two systems is seen in several sections.

The Tournaisian and Visean comprise thick successions of calcareous rocks - oolites, crinoidal limestones, dolomites, calcarenites - which reach maximum thicknesses (5000 ft; 1525 m) in S.W. Gower and southernmost Pembrokeshire. The only hints of unrest in these southern areas are the lagoonal character of the basal Visean sediments and the slight scouring of the underlying Caninia Oolite.

Eastwards the succession thins to a few hundred feet at the E. end of the Coalfield; dolomitisation is widespread and much of the Visean is missing. This was due to the uprise of the Usk Axis, which was to continue to exert an influence on sedimentation throughout the Carboniferous. West of Gower the sequence remains thick S. of the Ritec Fault but is thinner to the N.

Namurian uplift of St. George's Land to the N., and possibly the initiation of a Wales/Devon barrier in the S., caused flooding of the calcareous flats with sand and mud, and the deposition of the Millstone Grit. This Series, in the present region, is not marked by the thick, coarse sandstones which gave it its name in the Pennines, but it remains essentially a deltaic formation. Although showing considerable lateral variation, both in lithology and thickness, it is generally divisible into an upper and lower sandy group, separated by a shaley group, the Middle Shales. The upper sandstone is termed the Farewell Rock; only rarely are workable coals found beneath it. Again there is thinning towards the E.

The Coal Measures include a lower, shaley portion and an upper, mainly psammitic unit, containing the thick Pennant Sandstones. These are of sub-greywacke facies and mainly represent the molasse of a southerly, newly emergent, mountain chain; in the N., however, there was some deviation from that direction (Kelling, 1964). The rising Usk Axis also continued to exert an influence, as shown by breaks in the Coal Measure sequence, which became more marked towards the E. (Blundell, 1952). The Stephanian is missing.

The older strata are partly covered unconformably by Upper Triassic, non-marine beds (Keuper "Marls"), followed by marine Rhaetic and Lower Lias; basal breccia or conglomerate facies of the Triassic and shoreline facies of the marine beds are seen where the Mesozoic formations overlap onto the old land surfaces.

Tectonic Evolution

> Caledonian (sensu lato)
> Pre-Namurian (Sudetic)
> Post-Westphalian (Asturian)
> Alpine (sensu lato)
> Pliocene - uplift

Field descriptions of South Wales geology are given by Owen and Rhodes (1960), Owen et al., (1965) and Anderson (1971).

The Precambrian outcrops are too limited to provide evidence of Precambrian events. There is no evidence of Caledonian folding comparable to that of the Welsh Block (4.2). Mid-Lower Palaeozoic uplift is shown, for example, by the occurrence of Wenlock strata on Llanvirn in South Pembrokeshire and of Llandovery on Precambrian N. of Milford Haven. Pre-Lower Devonian disturbance also occurred, for this formation transgresses Silurian onto Ordovician when traced W. across Carmarthenshire. On the other hand, round the Usk Inlier there is no break between the Silurian and Lower Old Red Sandstone. Mid-Devonian, i.e. Late Caledonian, uplift took place but there was little or no folding.

The dominant structure of South Wales is the E.-W. downfold of the coalfield produced by post-Westphalian compression. This was preceded, however, by movements throughout the Carboniferous, the most important being Sudetic uplift which is marked by a major break at the base of the Namurian. The uplift of the Usk Axis played a major role. The South Wales Coalfield is a major downfold, which includes a number of minor folds, extending for 60 miles from the Burry Estuary in the W. to Pontypool in the E. East-west faults sometimes replace the fold axes. East of the Llanwonno Fault, the Pontypridd Anticline is comparatively gentle and symmetrical, but further W. the southern limb steepens to $50°$.

The Coalfield is extensively faulted, the main fault directions being N.N.W.-S.S.E., N.E.-S.W. and E.-W. A classification has been given by Woodland and Evans (1964) for the faults in the Pontypridd-Maesteg districts:
(i)   strike-thrusts, confined to the South Crop;
(ii)   normal strike-faults;
(iii) dip or cross-faults;
(iv)   incompetence faults.

The anthracite belt of the Gwendraeth Valley and Cynheidre region is notorious for thrust structures of Group (i), as also are the Cross-Hands and Ammanford areas (Trotter, 1947).

The Neath and Swansea Valley disturbances trend N.E. across the central portion of the Coalfield and then cut across the Old Red Sandstone belt into Herefordshire, the Neath Disturbance extending as far as the Malvern-Abberley line (6.4). These two disturbances involve narrow belts of faulting and of impersistent but often locally sharp folding. The main fractures in both disturbances are single wrench-faults with sinistral displacements of about three-quarters of a mile. Owen (1954) has suggested that the anomalous trend of the Neath Disturbance is due to basement control by a deep-seated fracture belt which resolved the N.N.E.-S.S.W. Hercynian compression along its length.

South of the E. part of the Coalfield the W.-pitching Cardiff Anticline brings up Silurian in the Rumney Inlier, Cardiff and Lower Old Red Sandstone in the Vale of Glamorgan. Further W. folding S. of the Coalfield becomes more intense in the Gower Peninsula where major anticlines bring up the Old Red Sandstone. The limbs of the folds are frequently steep, and overturning occurs on the northern limb of the Langland Anticline.

Thrusts trend parallel to the fold axes and include the Cefn Bryn, Port Eynon and Bishopston fractures. They dip southwards, a notable exception being the Caswell Thrust which dips northwards at $50°$ to $60°$. A large number of cross-faults trend N.N.E.-S.S.W., N.-S. and N.N.W.-S.S.E.. At first glance it seems that horizontal movement along these dip faults post-dated completion of the folds but in detail it can be seen that the fold intensity usually differs across the fracture, e.g. the Oxwich Point Anticline (see George, 1940, 175). George has clearly demonstrated that the folding, thrusting and cross-faulting were more or less contemporaneous and marked contrasts in structure occur across many of the cross-faults.

The Pembrokeshire Coalfield is a much narrower belt with a relatively gentle northern flank, but with a much contorted and thrust central and especially southern margin. The majority of the anticlines have steeper (even overturned) northern limbs and most of the thrusts dip southwards at $20°$ to $50°$. Many thrusts occur along coals and fireclays and some thrusts are themselves folded. The intense folding may be due to Hercynian push against the Welsh Lower Palaeozoic Block. Along the Johnstone Thrust, on the S. boundary of the Coalfield, Precambrian or Silurian is brought against the Coal Measures. This and other faults and sharp folds, according to one interpretation (5.11), may mark the Hercynian front. The Ritec Fault along Milford Haven is also possibly associated with the front, but it follows a deep structure as it influenced sedimentation from the Silurian onwards. Eastwards it may continue as the Swansea Valley and Neath Disturbance (Owen, 1954). S. of the Ritec Fault, in South Pembrokeshire, sharp, $W.10°N.$ trending folds, commonly with vertical limbs, affect Ordovician to Lower Carboniferous strata.

Uplift and deep erosion preceded the deposition of the Triassic/Rhaetic/Liassic succession now virtually confined to the S.E. of the region. These lie in broad folds well seen in the coastal cliffs W. of Cardiff. They are also much affected by faults, the majority of which trend E.S.E. or E. Normal faulting is common but some are reversed, e.g. at Southerndown. These post-Liassic folds and faults could have been formed by any Alpine phase but they are usually dated as Miocene like those of Southern England (8.8). They are certainly older than the 200 ft (60 m) coastal platform regarded, like higher platforms of the region, as due to Pliocene uplift.

6.6 The Bristol and Forest of Dean Coalfields and their Margins (Fig. 6.6)

This is a triangular area with a base along the S. margin of the Mendip Hills and an apex at the N. end of the Forest of Dean Coalfield, which lies N. of the Severn Estuary. The Mendip Hills rise to 1068 ft (325 m), but most of the region consists of low hills and flattish ground with rather sparse outcrops.

The succession is much the same as that in South Wales (6.5). There is, however, a Cambrian outcrop on the Malvern Axis, S. of the Severn; on the other hand, the Precambrian and Ordovician do not appear. The Silurian, Old Red Sandstone and Lower Carboniferous all contain basic lavas. Some excursions are described by Cowie et al., (1965).

Tectonic Evolution

Compared with those of the Welsh Coalfield the structures are more complex and show a wider variety of trends. This is partly due to the greater proximity of the region to the Hercynian front and partly due to the influence of two ancient lineaments. These are the Bath Axis, the southward continuation of the Malvern Axis (6.4) and the caledonoid Lower Severn Axis, which branches S.W. off the Malvern Axis where it crosses the Severn. In the Bath Axis the Bath hot springs, famous since Roman times, are of deep-seated origin. They have a temperature of $120°F$ and both water and natural gas are radioactive; the CO_2 free gas contains 13 cc of rare gases per litre.

The thick Cambrian (Tremadoc) shales and thin sandstones outcropping at Tortworth, just S. of the junction of the Axes, are overlain unconformably by the Llandovery Series followed conformably by the rest of the Silurian and the Downtonian. This suggests movements during the Ordovician. Mid-Devonian disturbances were more powerful than in South Wales, for in the Chepstow area N. of the Severn and in the Clevedon area to the S. the Upper Old Red Sandstone rests with distinct unconformity on the Lower.

According to one interpretation the Hercynian front passes S. of the Mendips and, therefore, S. of the Dinantian carbonate facies. Viewed broadly the Mendips form part of the generally E.-W. Mendip Axis which, as it crosses the Bath Axis, is deflected to the S. On the other hand, (for discussion see 5.11), the Hercynian front may pass eastwards from Pembrokeshire (6.5) under the Bristol Channel and then be diverted N.E. (and perhaps broken up as in South-East Ireland, 5.12) by the Lower Severn and Malvern/Bath Axes.

Marked Sudetic movements occurred in the region, forming roughly N.-S. folds in the Forest of Dean and N.-E. folds S. of the Severn. In both the Forest of Dean and Clevedon districts Coal Measures rest on Lower Carboniferous or Old Red Sandstone. The occurrence of sandstone in the upper part of the Dinantian is also evidence of early Carboniferous unrest.

Four major E.-W. periclines bringing up Lower Carboniferous and Old Red Sandstone form the Mendip Hills. From W. to E. they are the Blackdown, North Hill, Pen Hill and Beacon Hill upfolds; they are arranged <u>en echelon,</u> being offset further N. in a westerly direction. The two closest folds are the central pair, the intervening syncline being virtually cut out by the Emborough Thrust. All four folds are asymmetrical, the northern limbs being steeper. Much thrusting aligned N.W.-S.E. occurs in the Westbury area, S.W. of the North Hill fold, as, for example, the Ebbor Thrust. The Blackdown upfold is relatively free of cross-faulting but such fractures increase in frequency eastwards across the Mendip belt. Two important cross-faults cut across the North Hill Pericline. The western (Stock Hill) fracture is clearly pre-Triassic, whilst the eastern (Biddle) fault has moved again in post-Rhaetic times. Strike thrusts, especially on the southern limbs of the upfolds, have frequently originated at the lubricant horizon of the Lower Limestone Shales.

The echelon pattern of the Mendip flexures is obviously the result of a northward push against the N.-S. aligned bulwark of the Bristol-Radstock downfold and its parallel eastern upwarp, the Bath Axis. This resistance held back the eastern end of the Mendip ripple, allowing the Blackdown Pericline to form freely to the W. With increasing northward push, cross-shearing accompanied by forward thrusting took place. The most severe final results of the increasing resistance in the E. were (a) the thrusting of the Pen Hill fold tightly against the North Hill Pericline and (b) the heaving forward of klippen of Carboniferous Limestone and Millstone Grit onto vertical, overturned and highly contorted Coal Measures on the N. side of the Beacon Hill fold.

Further N. the influence of the old structures becomes more marked, and in the Bristol Coalfield folds trend E., N.E. and N. One of the most remarkable structures is the Kingswood Anticline in which disharmonic folding, affecting the incompetent Lower Coal Series, is accompanied by thrusts from both N. and S. It is probable that the more competent Carboniferous Limestone is only gently arched. This complex structure is thought to have been formed by the northwards Hercynian push into the constriction between the converging Bath and Lower Severn Axes (Kellaway & Welch, 1948). To the N.W. the Westbury-on-Trym pericline with Old Red Sandstone in its core has an almost vertical N.W. limb, beyond which lies the shallow syncline of the Avonmouth coal basin. The S.E. flank of the pericline includes the classic Avonian section of the Lower Carboniferous.

North of Bristol there is a major N.-S. trending anticline on the Malvern Axis, with Coal Measures to the S., the direction of pitch. <u>En echelon</u> to the W., N. of the Severn, the Forest of Dean underwent further post-Westphalian N.-S. folding.

The Malvern Axis continued to exert an influence during the Alpine movements. The gently dipping Mesozoic strata have a northerly strike, and a shallow N. by E. syncline with Lias in its core overlies part of the Hercynian anticline N. of Bristol, but with its axis offset to the W. of that of the older structure.

6.7 Central Ireland (Fig. 4.8)

Central Ireland is regarded as extending to the Highland Boundary Fracture-zone (3.3) on its (partially hidden) course across Ireland, excluding the Irish "Southern Uplands" (4.4) and Connemara (3.4). In the S. it extends to the Hercynian front (5.12). Apart from a coastal stretch near Dublin, on the E. the region comes against the Lower Palaeozoic and granite of South-East Ireland (4.5).

In the W. fine coastal exposures overlook the Atlantic. Good inland exposures also occur on the higher ground but much of the region is between only 200 and 400 ft (60-120 m) above sea-level (the Central Irish Plain) and is deeply mantled in glacial deposits. Hills and mountains, mainly eroded in anticlinal inliers of Old Red Sandstone and Silurian, are more common in the S. and rise to 3127 ft (952 m) at Brandon Mountain, N. of Dingle Bay.

The succession is as follows :-

     ? Oligocene
     Eocene (lavas)
     Cretaceous
     Jurassic (Lias)
     Triassic
     Permian
     Coal Measures
     Millstone Grit
     Carboniferous Limestone
     Old Red Sandstone
     Ordovician
   ? Dalradian

The oldest rocks occur in the N.E. of the region, in the Tyrone Inlier. These are biotite-schists which become of higher grade, with sillimanite, towards the S.W., though this is probably due to thermal metamorphism by later acid intrusions.

The schists could be Dalradian, but they are of higher regional grade than Dalradian rocks exposed a short distance away N. of the Highland Boundary Fracture-zone (3.3). The schists, therefore, may be older Precambrian rocks (Charlesworth, 1963, 25).

The metasediments are framed, above a thrust, by the Tyrone Igneous Series, consisting of spilites overlain by tuffs, cherts, black-shales and ferruginous tuffs. These are certainly Ordovician, but fossils are poorly preserved and age estimates range from Llanvirn to Lower Caradoc.

To the N. of Dublin, near Balbriggan and Portrane, Caradoc slates and greywackes occur, associated at Portrane with limestones and volcanics. Southwest of Dublin, at the Chair of Kildare, similar strata are underlain by Llanvirn gritty shales. Caradoc shales, cherts and greywackes also occur in the Slieve Aughty and Slieve Bernagh areas N. of Limerick.

The major parts of the Silurian outcrops in Central Ireland comprise rocks of Llandovery age. They occur in the inliers of Balbriggan, the Galtees, Slieve Bernagh and Slieve Bloom. The rocks include red, grey and green greywackes, mudstones, slates, flags and shelly conglomerates. The unfossiliferous slates (good enough, in such places as the Lingam Valley, to be used for roofing), grits and greywackes of Slievenaman, the Comeragh Mountains and Slieve Aughty are probably also of Llandovery age. In the Chair of Kildare Inlier, Llandovery red and black shales may follow Ashgillian Beds with little stratigraphical break. This may be true also at Balbriggan and Portrane (McKerrow, 1962). Wenlock strata are found in the Slieve Felim and Silvermines areas, the Keeper Hills and on Devil's Bit Mountain. The rocks consist of dark grey and blue mudstones, siltstones and interbedded greywackes. Wenlock flags occur in the Cratloe Hills, whilst highly cleaved slates with greywackes in the Galtee Mountains are probably of basal Wenlock age. The Wenlock-Ludlow succession on Devil's Bit Mountain is over 8000 ft (2440 m) thick, whilst that of the Slieve Bernagh Inlier includes spectacular slump-conglomerates, one formation being over 700 ft (214 m) thick.

In the Dingle Peninsula the Silurian is followed unconformably by conglomerates, grits, shales, etc., identified as Downtonian. These are followed, with major unconformity, by fluviatile beds which are probably Upper Old Red Sandstone. In fact, McKerrow (1962) suggests that the Dingle Beds are the only Lower Old Red Beds in the southern part of the region.

The Old Red Sandstone appearing in the southern inliers comprises red, purple, green and grey sandstones, shales, flags, and occasionally thick conglomerates (even of cobble grade) and breccias. The succession is extremely thick (over 10,000 ft, 3050 m) in the Comeragh-Knockmealdown boundary region, and a sub-caledonoid trough of maximum thickness probably then extends across County Cork towards Kenmare, near which locality the Old Red Sandstone sequence reaches over 15,000 ft (4600 m) in thickness. A marked thinning takes place, however, both northwards and north-eastwards; around the Slieve Felim and Slieve Bernagh inliers only about 1000 ft (305 m) of rocks represent the Devonian System. In the N., the Curlew Mountains pericline (Charlesworth, 1963, 195) brings up fluviatile sediments and volcanics of probable Lower Old Red age. The Fintona beds, forming a large outcrop immediately S. of the Highland Boundary Fracture-zone, are of the same facies as the Old Red Sandstone in the same structural position in the Midland Valley of Scotland (6.2), but it is not clear to which division they belong.

The Carboniferous Limestone covers at least two-thirds of the surface of the region. Its thickness over Central Ireland is about 3000 ft (915 m), thinning towards the margins of the Leinster and Longford-Down massifs but swelling to almost 4000 ft (1220 m) in the intervening Dublin Basin. The most important single element of the Dinantian succession in the S. part of Central Ireland is the great sheet reef, which extends as far N. as County Clare in the W. and County Kildare in the E., an area of nearly 3000 square miles. This massive reef, nearly 2000 ft (610 m) thick in the Cork district, belongs to the lowest Visean and consists of polyzoan and algal fronds, often in position of growth. In the Limerick district the Dinantian contains two basaltic horizons. Fragments of Leinster granite and mica-schist in the Upper limestone of the Clondalkin district point to active fault scarps along the N. edge of the Leinster Massif at that time.

At the top of the Dinantian sequences occur black Pendleside-type shales and thin limestones, and on the S. side of the Balbriggan Massif these are followed by sandstones and shales (Smyth, 1950). In South Clare, in the vicinity of the Shannon estuary, the Namurian is 3500 ft (1065 m) thick, comprising shales, flags, siltstones and sandstones and with many slump sheets and even sand volcanoes (Brindley & Gill, 1959). In North Clare the sequence thins appreciably. Similar thinning occurs southwards from the Shannon estuary (Hodson & Lewarne, 1961).

Coal Measures occur over an area of 150 square miles in the Leinster or Castlecomer Coalfield. This is an oval-shaped plateau with a central depression so that the Coalfield is a basin both structurally and topographically, the outer rim comprising tough Millstone Grits rising to between 500 and 1000 ft (152-305 m) O.D. Three workable seams of anthracite occur in the lowest 800 ft (245 m) of the Westphalian succession. The other coalfields of Central Ireland are very much smaller, but again form ridges above the central plain.

In the extreme N.E. of the region an interrupted Permian/Mesozoic succession, overlain by the Antrim Tertiary basalt lavas, overlaps the Highland Border (3.3) onto the continuation of the Midland Valley of Scotland. The basalts are overlain by the Loch Neagh clays, shown in a bore to be over 1100 ft (335 m) thick. They may be Oligocene, although an age as late as the Pliocene has also been postulated.

Some excursions are described by Nevill (1965).

Tectonic Evolution

Caledonian
(?Mid-Ordovician
(Post-Silurian/Pre-Devonian
(Mid-Devonian

Hercynian
(Pre-Namurian (Sudetic)
(Post-Westphalian (Asturian)

Alpine (sensu lato).

Evidence for pre-Silurian folding is slight in Central Ireland, although movements are suggested by breaks in the Ordovician successions. The intense folding of the Caradoc limestones at Portrane may be tectonic, but is ascribed by Lamont (1938) to slumping due to seismic shocks. Caledonian folding certainly took place at the end of the Silurian, as shown by the break beneath the Downtonian in the Dingle Peninsula. The Mid-Devonian Caledonian phase was important, as shown by the Upper Old Red relationship to the Downtonian mentioned above and by the absence of the Lower Old Red from much of the region. The Caledonian trends movements produced N.E. or E.N.E. (towards the W.) anticlinoria and synclinoria, and axial plane cleavage was formed in places, e.g. in the Slieve Bernagh area; strike faults also occur. It is, however, sometimes difficult to differentiate some of the Caledonian structures from later Hercynian effects.

Breaks in the Dinantian/Namurian sequence show that some disturbances occurred in mid-Carboniferous times, but Sudetic movements were not important in Central Ireland. The main Hercynian compression was post-Westphalian and although it was regionally N.-S. its effects were influenced by several factors. Perhaps the most important of these was the N.E. to E.N.E. trend of the already existing Caledonian folds. Bordering massifs also played an important part, especially the Leinster Block to the E. with its large granite (4.5). The wedging of the Old Red Sandstone from about 15,000 ft (4600 m) in the S. to virtually nothing in the Dublin-Galway Bay areas, and then its increase in thickness again towards the Highland Border, also affected the fold styles. Proximity, in the S., to the Hercynian front is marked by increase in complexity.

The Hercynian folds of Central Ireland, when traced from W. to E., change from an E. by N. direction through N.E. to N.N.E. towards the W. edge of the Leinster Massif. In fact, in the Leinster Coalfield almost N.-S. folds appear. The folds then swing E. again, round the N. end of the Massif near Dublin, where they are numerous and sharp probably due to compression between the Leinster and "Southern Upland" blocks. In the N. of the region, around Athlone, the Carboniferous blanket was only gently rippled. This still applies in the N. of Clare; however, as the Upper Carboniferous sandstones and shales are traced towards the S.W. of the county the folds are sharp and dips moderate to steep.

In the Leinster Coalfield the folding is gentle, but in the Slieve Ardagh field, 16 miles (25.5 km) to the S.W., the folds are much sharper. In the Kanturk or North Cork field, just N. of the Hercynian front, the structure is complex and the Coal Measures are overthrust from the S. by older strata (Nevill, 1966). The seams are crushed and in places have been swelled tectonically to many times their true thickness.

Faulting in Central Ireland follows a number of directions, the longest fractures trending E.N.E. Faults striking N. are also common, particularly in the coalfields; in the Leinster Coalfield, for example, the Luggacurren Fault has an easterly downthrow of 700 ft (215 m).

Alpine deformations can be clearly distinguished from older structures only in the N.E., where they affect the Tertiary lavas, etc. Folding and faulting of this age in the Highland Border area have already been described (3.3). Extensive faulting of the Lough Neagh clays is evidence of movements well into the Tertiary.

REFERENCES

Anderson, E.M. 1951. *The Dynamics of Faulting and Dyke Formation in Britain.* 2nd ed., Edinburgh.

Anderson, J.G.C. 1971. The Cardiff District. *Geol. Assoc. Exc. Guide,* No. 16. 2nd (revised) ed.

Baker, J.W. 1971. The Proterozoic History of Southern Britain. *P.G.A.,* 82, 249-266.

Bassett, D.A. 1958. Geological Excursion Guide to the Glasgow District. Geol. Soc. Glasgow.

Bathurst, R.G.C. et al. 1965. Geology around the University Towns. *Exc. Guide, No. 6, Geol. Assoc.* (Liverpool).

Blundell, C.R.K. 1952. The Succession and Structure of the North-Eastern Area of the South Wales Coalfield. *Q.J.G.S.,* 107, 307-333.

Brindley, J.C. & Gill, W.D. 1959. Summer Field Meeting in South Ireland. P.G.A., 69, 244.

Broadhurst, F.M. et al. 1959. The Area around Manchester. Exc. Guide No. 7. Geol. Assoc.

Brooks, M. 1970. Pre-Llandovery tectonism and the Malvern Structure. P.G.A., 81, 249-268.

Butcher, N.E. 1962. The tectonic structure of the Malvern Hills. P.G.A., 73, 103-123.

Charlesworth, J.K. 1963. *Historical Geology of Ireland.* Edinburgh.

Cope, F.W. 1954. In: *The Coalfields of Great Britain,* (ed. A.E. Trueman), London.

Cope, F.W. 1963. Week-end Field Meeting in the Peak District of Derbyshire. P.G.A., 74, 91-96.

Cope, F.W. 1976. The Peak District of Derbyshire. Exc. Guide No. 26. Geol. Assoc. (2nd ed.).

Craig, G.Y. & Duff, P. McL.D. 1976. *The Geology of the Lothians and South-East Scotland.* Edinburgh.

Cowie, J.W. et al. 1965. Field Meeting in the Bristol District, North of the River Avon. P.G.A., 76, 261-267.

Curtis, M.L.K. 1968. The Tremadoc Rocks of the Tortworth Inlier, Gloucestershire. P.G.A., 79, 349-362.

Dunham, K.C. et al. 1965. Granite beneath Visean sediments with mineralisation at Rockhope, Northern Pennines. Q.J.G.S., 121, 383-427.

Dunham, K.C. 1967. Veins, Flats and Pipes in the Carboniferous Limestone of the English Pennines. Econ. Geol. Mono., 3, 201.

Dunham, K.C. 1973. A recent Deep Borehole near Eyam, Derbyshire. *Nat. Phys. Sci.,* 241, 84-85.

Dunham, K.C. 1974. Granite beneath the Pennines in North Yorkshire. Proc. Yorks. Geol. Soc., 40, 191-194.

Dunning, F.W. 1975. Precambrian craton of central England and the Welsh Borders. Spec. Rep. No. 6, Geol. Soc., 89-94.

Falcon, N.L. & Kent, P.E. 1960. Geological results of petroleum exploration in Britain, 1945-57. Mem. Geol. Soc. London, 2.

Ford, T.D. 1963. The Pre-Cambrian Fossils of Charnwood Forest. *Trans.* Leicester Lit. & Phil. Soc., 57, 57-62.

Garrett, P.A. et al. 1960. The Area around Birmingham. Exc. Guide No. 1. Geol. Assoc.

George, T.N. 1940. The structure of Gower. Q.J.G.S., 96, 131.

George, T.N. 1963. Tectonics and palaeogeography in northern England. Sci. Prog., 51, 52.

Hodson, F. & Lewarne, G.C. 1961. A Mid-Carboniferous (Namurian) Basin in parts of the counties of Limerick and Clare, Ireland. Q.J.G.S., 117, 307-333.
Johnson, G.A.L. 1973. The Durham Area. Exc. Guide No. 15. Geol. Assoc.
Kellaway, G.A. & Welch, F.B.A. 1948. British Regional Geology: Bristol and Gloucester District. M.G.S.
Kelling, G. 1964. Sediment Transport in part of the Lower Pennant Measures of South Wales. Devel. in Sedimentology, 1, 177-184.
Kennedy, W.Q. 1958. The tectonic evolution of the Midland Valley of Scotland. T.G.S.G., 23, 106-137.
Lamont, A. 1938. Contemporaneous slumping and other problems at Bray Series, Ordovician and Lower Carboniferous Horizons in County Dublin. P.R.I.A., 45B, 1-25.
Lapworth, C. 1882. The Girvan Succession. Q.J.G.S., 38, 357.
Le Bas, M.J. 1972. Caledonian igneous rocks beneath central and eastern England. Proc. Yorks. Geol. Soc., 39, 71-86.
Lees, G.M. & Taitt, A.H. 1946. The geological results of the search for oilfields in Britain. Q.J.G.S., 101, 255-317.
Macgregor, A.G. 1968. Fife and Angus Geology. Blackwood.
McKerrow, W.S. 1962. The chronology of Caledonian folding in the British Isles. Proc. Nat. Acad. Sci., 48, 1905.
Nevill, W.E. 1965. Summer Field Meeting in Southern Ireland, 1962. P.G.A., 76, 305-314.
Nevill, W.E. 1966. The Geology of the North Cork (Kanturk) Coalfield. Geol. Mag., 103, 423-431.
O'Nions, R.K. et al. 1973. New isotopic and stratigraphical evidence on the ages of the Ingletonian, probable Cambrian, of Northern England. Journ. Geol. Soc., 129, 445-452.
Owen, T.R. 1954. The structure of the Neath Disturbance, Bryniau Gleision and Glyn Neath, South Wales. Q.J.G.S., 109, 333-359.
Owen, T.R. 1958. The Armorican earth movements. In: The Upper Palaeozoic. 7th Inter. Univ. Geol. Congress (Swansea), 46.
Owen, T.R. & Rhodes, F.H.T. 1960. Geology around the University Towns: Swansea, South Wales. Exc. Guide No. 17. Geol. Assoc.
Owen, T.R. et al. 1965. Summer (1964) Field Meeting in South Wales. P.G.A., 76, 463-496.
Peach, B.N. & Horne, J. 1899. The Silurian Rocks of Britain, vol. 1, Scotland. M.G.S.
Penn, J.S.W. & French, J. 1971. The Malvern Hills. Exc. Guide No. 4. Geol. Assoc.
Robertson, T. 1933. Geology of the South Wales Coalfield, Part I. The geology of the country around Merthyr Tydfil. 2nd ed. M.G.S.
Shiels, K.A.G. 1964. The geological structure of north-east Northumberland. T.R.S.E., 65, 449-481.
Smyth, L.B. 1950. The Carboniferous System in North County Dublin. Q.J.G.S., 106, 105-295.
Squirrell, H.C. & Tucker, E.V. 1960. The Geology of the Woolhope Inlier, Herefordshire. Q.J.G.S., 116, 139-185.
Taylor, K. & Thomas, L.P. 1975. Geological Survey boreholes in south Brecknockshire and their bearing on the stratigraphy of the Upper Old Red Sandstone boundary east of the Afon Hepste. Bull. Geol. Surv. G.B., No. 54.
Tomkeieff, S.I. 1965. Cheviot Hills. Exc. Guide No. 37. Geol. Assoc.
Trotter, F.M. 1947. The structure of the Coal Measures in the Pontardawe - Ammanford area, South Wales. Q.J.G.S., 103, 89.

Trotter, F.M. 1954. In: The Coalfields of Great Britain, (ed. A.E. Trueman), London.
Williams, A. 1959. A structural history of the Girvan district, south-west Ayrshire. T.R.S.E., 63, 629.
Williams, A. 1962. The Barr and Lower Ardmillan Series (Caradoc) of the Girvan district, south-west Ayrshire. Mem. Geol. Soc. London, 3.
Woodland, A.W. & Evans, W.B. 1964. Geology of the South Wales Coalfield, Part IV. Pontypridd and Maesteg. 3rd ed. M.G.S.
Ziegler, A.M. et al. 1969. Correlation and environmental setting of the Skomer Volcanic Group, Pembrokeshire. P.G.A., 80, 409-439.

Chapter 7
ALPINE FOLD-BELTS

7.1   Introduction (Fig. 7.1) - The Western Alps

The Alps, the spectacular mountain heart of Western Europe, are, for many geologists, the crowning glory of European tectonics. Mont Blanc rises to 4810 m (15770 ft) and a number of other peaks exceed 4000 m. Precipitous, overdeepened valleys, the upper parts often occupied by glaciers, separate the mountains, and high rock walls (parois) frequently display magnificent structures.

Moreover, the grandeur of the scenery has been matched by the skill, thoroughness and sheer physical determination of several generations of geologists whose work has made the Alps the most intensely studied orogen in the world.

The Western Alps extend in a 600 km arc from the Riviera coast through south-eastern France, parts of northern Italy and Switzerland to Liechtenstein and the borders of Austria. Here they are tectonically overlapped by the Eastern Alps which form much of Austria, north-eastern Italy and Yugoslavia. Only the contacts of the Eastern Alps with the Western will be described.

Radiometric dating has shown that Precambrian, Caledonian, Hercynian and Alpine crystalline rocks occur in the Western Alps. Unaltered or slightly metamorphosed sediments range from the Carboniferous (some Silurian has been identified on the Mediterranean coast) to the Pliocene. The Mesozoic and Tertiary sediments show striking variation in facies, related to palaeogeographical changes, both in space and time, in the Tethys geosyncline and its margins.

Although there are marked differences of opinion regarding the dating of the Alpine movements, it is accepted that the region has been tectonically active, to a greater or lesser extent, from the Triassic to the Pliocene and that vertical movements have continued up to the present. It is also clear that certain phases affected some segments more strongly than others. The following is a broad summary of the structure-building events.

```
                    Late Precambrian (inferred from radiometric dating).
                    Caledonian (inferred from radiometric dating).

                  ( Early Westphalian (Sudetic).
   HERCYNIAN      ( Post-Westphalian   -  minor folding.
                  ( Post-Stephanian    -  minor folding.

                  ( Middle Trias  -  vertical movements;  initiation of Briançon
                  (                   and Dolin geanticlines.
                  ( Lias  -  vertical movements;  further geanticlinal or
   ALPINE         (           platform development.
   (sensu         ( Middle Cretaceous  -  folding, particularly in Dauphiné.
    lato)         ( Late Cretaceous/Eocene  -  folding, including nappe formation.
```

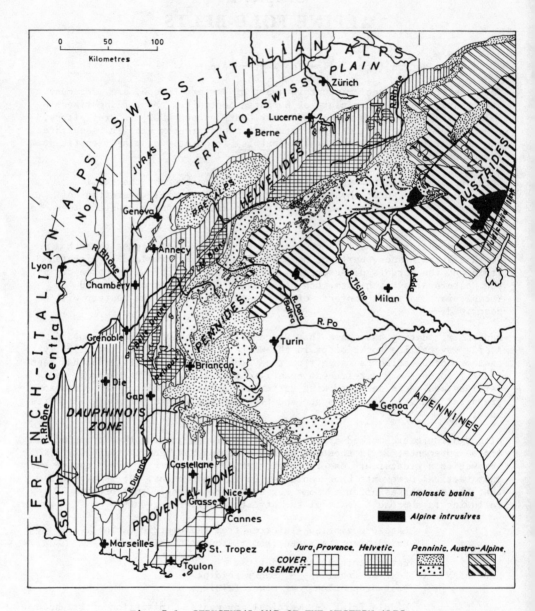

Fig. 7.1 STRUCTURAL MAP OF THE WESTERN ALPS
(after Debelmas and Lemoine 1970)

(Upper Eocene/?early Oligocene - folding, particularly in
(Provence.
(Lower Oligocene - major nappe formation, and later uplift,
(in Internal Zones.
(Late Miocene (Pontian)/Pliocene - second major fold period,
(probably general, including main folding
(of External Zones. Major vertical move-
(ments including uplift of External Zones
(and subsidence of molasse basins and of
(Mediterranean.

For long, Alpine structures were entirely attributed to horizontal crustal compression essentially due to the approach of the African and European shields or plates and to vertical crustal displacement; strike-slip or transform faulting was also recognised.

The occurrence of major uplift makes it possible for gravity tectonics to have operated, for example the Embrunais Nappes (7.12) may have glided westwards following the Lower Oligocene uplift of the Piémont Zone, and the Pre-Alps (7.4) have slid northwards following uplift of the External Zones and depression of the Swiss Plain. The spectacular structures within the Helvetic Nappes have been attributed to "cascade folding". Gravity tectonics do not preclude nappe tectonics but mean that detached portions of nappes need never have been continuous with the main nappe far behind it - thus removing a major difficulty of the nappe hypothesis.

The role of plate movements in Alpine tectonics is summarised in Chapter 10. The Western Alps (Gwinner, 1971) are divided into the Swiss-Italian Alps and the Franco-Italian Alps by a line roughly from Geneva to Turin, partly coinciding with the Franco-Swiss border (Debelmas and Lemoine, 1970). The separation is not tectonic, nor is there any dramatic change across the line. Nevertheless, major elements of these two segments show marked contrast in facies and in structure. The main structural units in segments of the Western Alpine Arc are given below.

SWISS-ITALIAN ALPS	FRANCO-ITALIAN ALPS		
	NORTHERN	CENTRAL	SOUTHERN
Juras	Juras		Dauphinois, Maritime Alps and Provençal Chains
Franco-Swiss Plain	Franco-Swiss Plain	Dauphinois Chains	
Helvetides	Helvetides		
Ultra-Helvetides	Ultra-Helvetides	Ultra-Dauphinois	
Pre-Alps and other Klippes	Annes and Sulens Klippes		
Hercynian* Massifs	Hercynian* Massifs	Hercynian* Massifs	Hercynian* Massifs
Pennides	Pennides	Pennides	
(Tectonic contact with Eastern Alps)		(Overlap of Molasse)	

* also known as External Massifs.

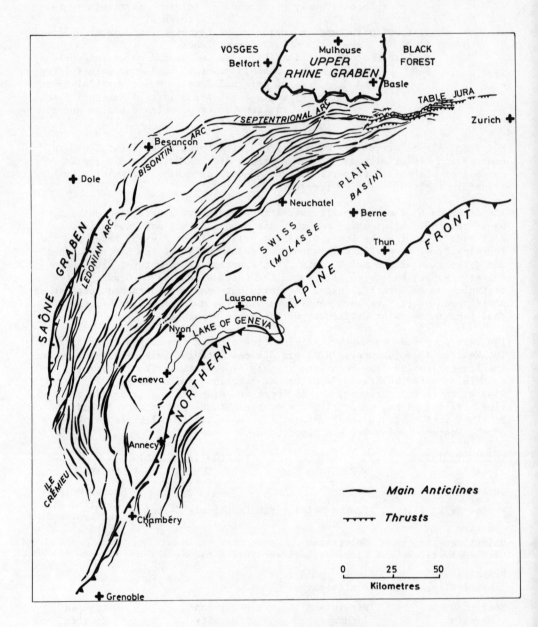

Fig. 7.2 STRUCTURAL MAP OF JURA MOUNTAINS

The Pennides constitute the Internal Zone, the other units the External
Zone; the Pre-Alps and other klippes are outliers of the Internal Zone, or
parts of the Eastern Alps, resting tectonically on the External Zone.
Excursions in the Swiss Alps are described in the guide published in 1967
by the Société Géologique Suisse, in the Jura by Chauve (1975) and in the
French Alps by Campredon and Boucarut (1975), by Debelmas (1970) and by
Gouvernet et al. (1971).

As the Juras and the Franco-Swiss Plain show only minor differences between
the Swiss and French segments they will be described as a whole (7.2 and
7.3).

7.2   The Jura Mountains (Figs. 7.2 and 7.3)

The Jura folds emerge from beneath the Tertiary of the South German Plain
about 20 km N.W. of Zurich and continue for 300 km through $90°$ of arc to
Chambery, reaching a maximum width of 70 km.

The Jura Mountains gave their name to the Jurassic system and are, in fact,
largely composed of limestones, etc., of that age. The stratigraphical
succession ranges downwards to the Middle Trias and upwards to the Miocene.

The Juras also gave their name to the Jura type of topography - anticlinal
ridges and synclinal valleys. This is substantially true, but is not the
whole picture. Where erosion has broken through the hard Jurassic limestones of the steeper anticlines into softer Jurassic shales and marls
beneath, reversal of topography has taken place forming strike valleys
(Fr. combes). Narrow transverse valleys (Fr. cluses; G. Klus) deeply
penetrate the Juras and provide routes for transport.

The main or inner arc of the Juras forms the Folded Jura or High Chain,
rising to just over 1500 m. In the eastern part of the Chain, S. of the
Black Forest, the Folded Jura are bordered to the N. by the Table Jura.
These consist of fault-separated and sometimes tilted blocks, mainly of
Mesozoic. The boundary with the Folded Jura is sharp and usually consists
of a thrust.

Further W., however, S. of the Upper Rhine graben and of the Vosges, the
Folded Jura pass gradually to the N.W. into the Plateau Jura. The latter
are formed of wide synclinal blocks separated by narrow fold-belts. Of
these fold-belts, the most marked from E. to W. are the Arc Septentrional,
the Arc Bisontin and the Arc Ledonien. The development of the much more
open structures of the Plateau Jura, compared with those of the Table Jura,
is due to the freedom of the folds to expand into the Upper Rhine and Saône
graben and into the corridor of Tertiary which joins the two graben S. of
the Vosges, forming the Belfort Trench or Burgundian Gate.

Boring (Lienhardt, 1962) immediately E. of the Saône graben, S.W. of
Besançon, to nearly 1000 m, has shown that the Mesozoic of the Arc Ledonien
has been pushed westwards along a composite thrust over a basement consisting of thick Permian and Upper Carboniferous sediments, including a limnic
coal basin of Stephanian age with several coal beds, resting on mica-
schists, gneisses and granites.

The southernmost outpost of the Plateau Jura forms the Ile-de-Crémieu in a

big bend of the River Rhone, just E. of its confluence with the Ain. At the
S.W. corner of this block there is a small exposure of basement metamorphic
rocks (Ager et al., 1963, 329). These are in faulted contact with the
Mesozoic rocks that make up the Ile, which form a gentle south-easterly dip-
slope away from the prominent fault-scarp of their north-western boundary.

The Folded Jura consist of a bundle (faisceau hélvetique) of sharp anticlines and synclines, the strike of which swings round from E., S. of the
Black Forest, to S., near Chambéry. Many of the folds are broken by
thrusts. The thrusts and overfolds are often directed towards the N.W.,
but this is not the whole story as there is overriding towards the S.,
particularly in the southern part of the Chain. The folds are well
displayed between Neuchatel and the Rhine graben where some ten major anticlines can be distinguished. In the French Jura, between the Ile-de-Crémieu
and the Lac de Bourget, there are five strongly developed anticlines with
intervening synclines. The lake occupies a deep synclinal valley separated
to the E. by another anticline from the Franco-Swiss Plain.

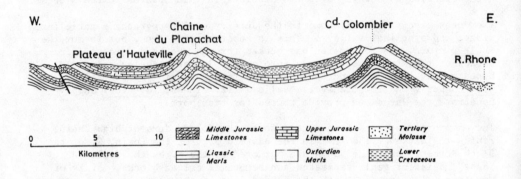

Fig. 7.3 SECTION OF SOUTH-WESTERN PART OF JURA MOUNTAINS

The folds of the Jura are regarded as "epidermal" and to have formed above
a décollement lubricated by evaporites (gypsum/anhydrite/salt) in the
Muschelkalk; this view mainly derives from Buxtorf's (1908, 1916) studies
of the railway tunnels. Later work has demonstrated the occurrence of
décollements at higher levels in the Jurassic. Boring has shown that the
Trias underneath the Jura Mountains is generally thicker than 500 m.,
reaching 1300 m in the centre of the Folded Jura. The development of salt
or other evaporites has taken place in both the Muschelkalk and the Keuper.
Moreover, the area of evaporite development almost exactly coincides with
the area of the Jura Mountains, which could lead to the conclusion that
where there are no evaporites there is no folding.

Some of the folds are of box or mushroom form. These are probably due to
diapiric folding, although a possibility is gravity cascading. Box folds

also occur in the southern French Jura (Ager et al., 1963). Breaks in the Jurassic sequence provide evidence of intra-Jurassic movement, including the uprise of anticlines which interrupted sedimentation.

Another major tectonic element of the Folded Jura are the wrench faults, at angles of from 20° to 40° with the folds, along which the eastern side is usually displaced northwards. Slips of up to 10 km (on the Vallorbe-Pontarlier fault) have been claimed. More recent research has, however, suggested that structures cannot be reliably matched across the faults and that they are really fault zones separating sectors within which folding took place separately (e.g. Aubert, 1959).

The structures of the Juras can be readily seen by following the routes traversing the Chain, for example, in the Swiss sector, the roads from Basel to Olten and from Le Socle to Neuchatel. In the southern French sector (Ager et al., 1963, 353) the cluse followed by the road and railway from Ambérieu to Culoz, and probably a former course of the Rhône, reveals spectacular folds and reversed faults.

7.3 The Franco-Swiss Plain or Molasse Basin (Fig. 7.1)

This is the lowland or lake district of Switzerland, containing the greater part of the country's industry and population. During the uprise of the mountains this was the site of the "foredeep", covered at times by the sea and at times by lakes, that received the products of the erosion of the mountains, mostly gravel and sand; some of the resulting conglomerates are known as nagelfluh. In the N. the pebbles have been derived from the Juras and the Black Forest, on the S. from the high Alps, recalling the Old Red Sandstone of the Midland Valley of Scotland (6.2). The molasse sediments are mainly of Upper Oligocene and Miocene age and are of both marine and freshwater origin.

The molasse is not without structure. An anticline has been traced for about 320 km through almost the full length of the Franco-Swiss Plain. South of Geneva this brings up the Upper Jurassic and Cretaceous in the Monts Salève, broken by a sinistral transcurrent fault at Cruseilles.

The softer molasse sediments are not frequently exposed; on many parts of the Franco-Swiss Plain there is a heavy cover of glacial deposits. The sandstone and conglomerates, however, can often be seen in scarp faces and cuttings near the Swiss cities, such as Berne and Lausanne.

7.4 The Swiss-Italian Alps - The Pre-Alps (Figs. 7.4 and 7.5)

The Pre-Alps form one of the largest and most intensively studied "tectonic outliers" in the world. Controversy polarised between the view that their upper nappes travelled over the Western Alps from the southern part of the Tethys - "a part of Africa resting on Europe" - and the "solution courte", favoured especially by French geologists (e.g. Trumpy, 1955), that at furthest the highest of these nappes came from the lowest structure of the Austrides.

The Pre-Alps occur in two major lobes, the Romande Pre-Alps to the N.E., extending from the Lake of Thun to the Lake of Geneva, and the Chablais Pre-Alps between the Lake and the R. Arve. Tectonically these lobes rest on

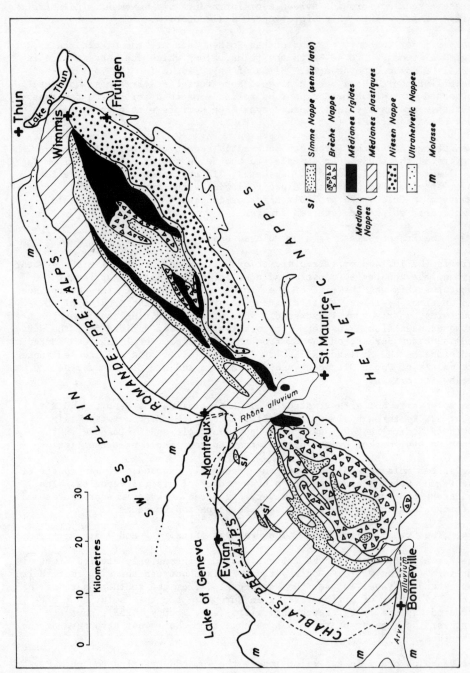

Fig. 7.4 STRUCTURAL MAP OF PRE-ALPS

the Helvetides and overlap onto the Swiss Plain.

The Pre-Alps lie in a major tectonic depression containing the following structural units:-

    Simme Nappe   (Simme Nappe (sensu stricto)
    (sensu lato)   (Plattenflysch Nappe
            (Gets Nappe

        Breche Nappe

    Median Pre-Alps (Medianes rigides
    Nappe   (Medianes plastiques

        Niesen Nappe

       Ultrahelvetic Nappes

The Simme Nappe (sensu lato) probably comes from the Piémont Zone of the Pennides, a view supported by the abundancy of ophiolites in the Gets Nappe and by the facies of the Plattenflysch Nappe, which is made up of the Helminthoid Flysch of Upper Cretaceous age. The Simme Nappe (sensu stricto), on the other hand, consists of Upper Cretaceous Flysch which may have come from the Sesia and Canavese Zones of the frontal parts of the Austrides.

The Brèche Nappe is made up almost entirely of Jurassic breccias and is thought to have originated from the southern border of the St. Bernard geanticline.

The Median Pre-Alps Nappe contains two sub-units. The Médianes rigides, consisting of thick Triassic breccias followed by massive calc-dolomites, and the Médianes plastiques, occurring more on the N.W. side of the Pre-Alps and consisting of Rhaetic followed by well-bedded Liassic Limestone, succeeded by Jurassic and Cretaceous Limestone and marls. Middle Cretaceous is absent in the Median Pre-Alps. The facies of the Médianes rigides and the Médianes plastiques correspond respectively to those of the Briançonnais and sub-Briançonnais Zones of the Franco-Italian Alps.

The Niesen Nappe consists of 2000 m of Niesen Flysch, mainly of Late Cretaceous age, consisting of rapid alternations of dark grey shales and greywacke-like sandstones with lenticular beds of coarse grit. It is probably derived from the Valaisian Zone and may perhaps be correlated with the Tarentaise Breccia Nappe at the front of the Pennides.

The Ultrahelvetic Nappes of the Pre-Alps, sometimes referred to as the External and Internal Pre-Alps or Col Nappes, were derived from the Ultrahelvetic trough on the S. margin of the Helvetides, and have not, therefore, travelled as far as the other Pre-Alp nappes. They are noteworthy for the presence of the Wildflysch, consisting of Tertiary shales, sandstones, quartzites and limestones, with exotic blocks of granite and other crystalline rocks, believed to have been derived from the Ultrahelvetic Cordillera immediately to the S. of the trough.

Parts of the Pre-Alps can be readily studied from the E. end of the Lake of Geneva. There are also easily accessible sections of their N.E. end near Wimmis and good sections in the flysch, which forms the great pyramid of the

Niesen above the Lake of Thun, can be seen near Frutigen.

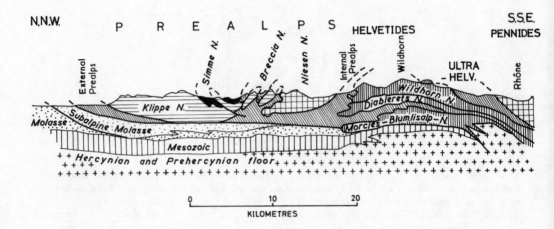

Fig. 7.5 SECTION OF HELVETIDES AND PRE-ALPS IN SWITZERLAND
(after Gwinner 1971)

7.5 The Swiss-Italian Alps - The Helvetides and Ultrahelvetides
(Figs. 7.5 and 7.6)

The Helvetides, in some respects the most spectacular, if not the highest, part of the Alps, are marked by jagged ridges overtopping high, near vertical rock faces, such as those around Kandersteg, in the Glarus and above the Wallensee. The cliff makers are massive limestones in the Lower Cretaceous, and, particularly in the S., similar limestones in the Upper Jurassic. In fact, the ranges are often referred to as the High Calcareous Alps. In Switzerland the Helvetides consist mainly of great décollement nappes, formed of epi-continental Mesozoic and early Tertiary stripped from its Hercynian basement. Around parts of this basement, particularly the Aar Massif, autochthonous sediments are still preserved. Involved with the Helvetic Nappes, and probably pushed over the Helvetic domain before the folding of the latter, are the Ultrahelvetic Nappes, the sediments of which were formed in deeper water between the Helvetic domain and an early southern cordillera. These consist mainly of flysch, including the Wildflysch, containing large exotic blocks derived from the cordillera.

The fronts of the Helvetic Nappes often show spectacular folds, sometimes described as "breakers at the nappe front". On the other hand, the main contacts between nappes are often remarkably smooth; for example, along the famous Glarus overthrust Permian red beds overlie Lower Tertiary flysch, the zone of disturbance being exceedingly thin, even of the order of 10 cm.

A classic section of the Helvetides occurs where the Rhône breaks through the Chain downstream from Martigny. In the Dents de Morcles, on the E. side

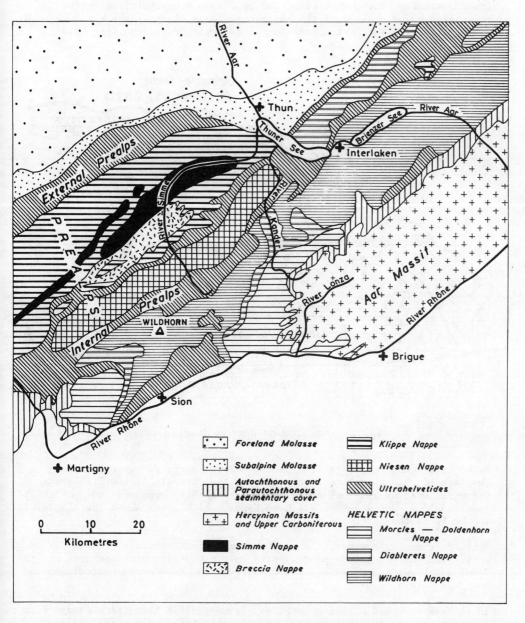

Fig. 7.6 STRUCTURE OF PART OF SWISS ALPS
(after Gwinner 1971)

of the valley, the flysch of the autochthon, owing to overlap, rests on
Upper Jurassic. The thrust is marked by a zone of mylonitised gneiss
considered to be a slice of the basement dragged along by the nappe. Above
comes a reversed succession from Eocene to Lower Cretaceous (Fig. 7.7).

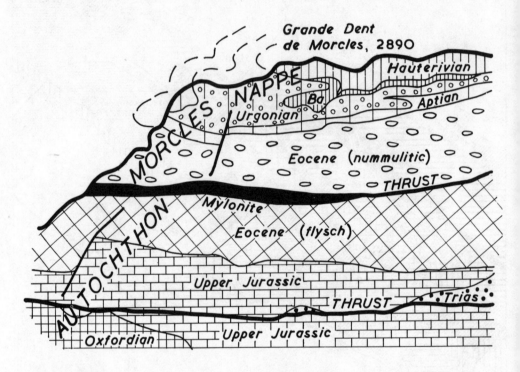

Fig. 7.7 SECTION OF DENT DE MORCLES CLIFF

As we have seen, the lower limb of the Morcles Nappe is reversed. The
Tertiary of the upper limb is overlain by the Jurassic of the Diablerets
Nappe followed by a normal succession so that this nappe does not possess a
reversed limb, nor does the Wildhorn Nappe, which comes above the Diablerets
to the E. in the direction of pitch.

Still further E., however, the pitch becomes south-westerly off the Aar
Massif and the Morcles Nappe reappears in the vicinity of Kandersteg and the
Gemmi Pass, resting on thin Mesozoic autochthon unconformably overlying the
Aar Massif (Fig. 7.8).

In the eastern part of the Helvetides (Trumpy, 1969) there is a change both
in structure and in stratigraphy. Essentially there is a single, large
Helvetic Nappe which includes over 100 m of Continental Permian at its base,
resting on the Glarus thrust. Parts of the Upper Mesozoic and Eocene,
however, have become detached to form secondary nappes; examples are the
Murtschen and Säntis Nappes, mainly made up of Upper Cretaceous, which form
the great cliffs on the N. side of the Wallensee. The Ultrahelvetic flysch
has been "diverticulated" into a number of thin nappes following plastic

incompetent horizons which act as levels of décollement. Some of these thin nappes are complexly entangled in the Helvetic Nappes.

The Helvetides are overlain by a few klippes which have been derived from the Pennides, the most famous being the Mythen, S. of Luzern, which consist of a Triassic and Cretaceous succession resting along a thrust on Tertiary.

Apart from the section already mentioned, downstream from Martigny, roads and paths near Kandersteg provide readily accessible and magnificent views of the Helvetic nappes and particularly of the "second order" folds. Further E., the valleys leading from Interlaken and Luzern, and the Linthal running S. from the Wallensee, are all easy means of access to the Helvetides.

7.6   The Swiss-Italian Alps - The Hercynian or External Massifs (Figs. 7.1 and 7.6)

Palaeozoic (and possibly Pre-Palaeozoic) (7.9) igneous and metamorphic rocks floored Helvetic sedimentation and have been up-thrust to form the External Massifs. The autochthonous cover of the Aar Massif, the more northerly of the two Massifs in Switzerland, is of Helvetic facies, that of the Gotthard of Ultrahelvetic or even Penninic facies.

The Aar Massif (Steck, 1968) has, broadly, a central core of granite and an envelope of gneiss and schist, at least partly of sedimentary origin. Distribution of the rock types is, however, complicated by thrusts, with overriding towards the N., which divide the Massif into wedges, some separated by very compressed sediments. Radiometric determinations give Hercynian ages but show an increase in the effects of Alpine metamorphism towards the S.E. The Aar Massif forms spectacular mountains often capped by jagged granite ridges, rising to over 3000 m, in the Bernese Oberland. Granite, gneiss and schist also occur in the Gotthard Massif, which is likewise divided into thrust wedges.

In the Upper Rhine Valley the Tavetscher Swischenmassiv intervenes between the Aar and Gotthard Massifs. This contains sub-vertical metasediments, such as phyllites, mica-schists and some green basic and ultrabasic schists, and is thought to be part of the zone of origin of the Helvetic Nappes. To have provided the great volume of these nappes it must have been originally wider and may have been narrowed by the approach of the Gotthard Massif to that of the Aar, possibly with consumption of some of the basement material in depth. Trumpy (1963) has suggested that this compression came after the stripping of the nappes from the area.

Parts of the Aar Massif can be readily studied S. of the Upper Rhone Valley, around the source of the Rhône at the Rhône glacier and near the Furka Pass. The Gotthard Massif outcrops between Andermatt and the pass which gives it its name, and the Tavetscher Swischenmassiv along the road from Disentis southwards towards the Leukmanier Pass.

7.7   The Swiss-Italian Alps - The Pennides (Figs. 7.1, 7.8 and 7.9)

The Pennides or Internal Zones are separated from the External Zones by the frontal Pennine Thrust, along which displacement of at least 40 km is claimed. The Pennides, particularly their gneissic cores, form very high

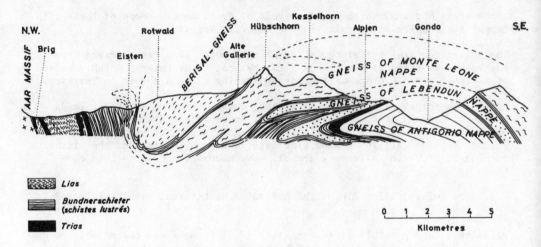

Fig. 7.8   SECTION OF SIMPLON-TICINO NAPPES
(after Gwinner 1971)

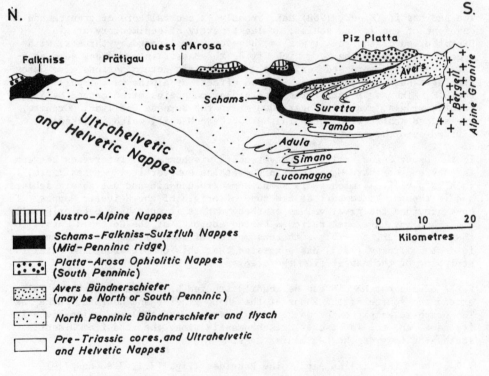

Fig. 7.9   SECTION OF THE GRISONS REGION
(after Debelmas and Lemoine 1970)

ground, including Monte Rosa, 4634 m and La Dent Blanche, 4356 m.

Three Lower Pennine Nappes, collectively referred to as the Simplon-Ticino Group, occur in a major culmination extending from the Ticino across the Simplon Pass, between Switzerland and Italy, to the Grisons. The Simplon-Ticino Nappes are characterised by the predominance of the gneissic cores of great recumbent folds, the envelopes of which are composed of relatively thin Mesozoic sediments which have undergone mesozonal metamorphism. The gneissic cores were earlier regarded as part of the Hercynian basement, but if so they have undergone complete Alpine recrystallisation and palingenesis, confirmed by radiometric dating (e.g. Jager et al., 1967). Wenk (1956) has put forward the view that they are syntectonic migmatites and gneissic granites formed in Tertiary times as the nappes advanced. This is also of significance with reference to the root-zone (see below).

To the W. of the culmination occur the three classic Pennine Nappes, the Grand St. Bernard, the Monte Rosa and the Dent Blanche, generally regarded as occurring in this structural order, although this is not completely certain. The cores of these nappes are certainly Hercynian in origin (e.g. the Casanna schists of the Grand St. Bernard, which have been formed from Upper Carboniferous, containing anthracite seams) but have also been affected by Alpine metamorphism. The envelopes are much thicker than those of the Lower Pennide Nappes, consisting of fairly thin Triassic followed by considerable thicknesses of schistes lustrés (or bunderschiefer), phyllites and mica-schists, thought to be mainly Liassic, deposited in the Valaisian palaeogeographic domain and containing ophiolites, probably part of the Mesozoic ocean crust. The Grand St. Bernard and Monte Rosa Nappes are closely involved with each other and, in places, difficult to distinguish. At their contact invaginated anticlines close southwards forming what is called the backward folding of the Mischabel, 4545 m, the highest mountain entirely in Switzerland. This is thought to be due to the Monte Rosa Nappe plunging into the back of the other nappe.

The Dent Blanche Nappe has a core of somewhat different character, consisting of the basic Valpelline and the more acid Arolla Series. A second order recumbent syncline within the Valpelline Series in its core forms the Matterhorn, 4477 m. Triassic and schistes lustrés surround the core and on Mont Dolin the Triassic, resting on the Arolla Series, is a coarse littoral deposit. The Dent Blanche Nappe is considered to be part of the Austrides which has overridden the Pennides.

The Upper Pennine Nappes also appear to the E. of the culmination where at least three nappes have been recognised, although correlation with those to the W. is controversial. Gneissic cores have thick envelopes of schistes lustrés (which range up to the Lower Cretaceous) seen in the Grisons basin. In the extreme E. of Switzerland and in Liechtenstein these nappes are overridden by the lowest nappes of the Eastern Alps. The Pennine facies, marking the continuation of the Pennine Nappes under the Austrides, reappears, however, further E. in the Engadine Window, and still further E. in the much larger window of the Hohe Tauern.

The southern margin of the Swiss-Italian Pennides is marked by the root-zone, where the gneissic masses separated by greatly compressed and tectonised metasediments become progressively vertical. Wenk (1956) put forward the view that these greatly compressed, steep gneisses represent magmatic domes which became bent over to form the cores of the Simplon-Ticino Nappes. This root-zone is sharply separated along the Insubric Line from the Southern Alpine or Austride basement rocks to the S. Within less than 100 m there is a contrast in Rb/Sr ages for micas of 11-25 m.y. to the N., with 300 m.y. to the S. Austride gneisses have clearly escaped Alpine influence (Jager et al., 1967) and the Insubric Line must, therefore, be a major dislocation, probably transcurrent (see also 10.2).

Along part of the southern margin of the Pennides the overfolding is directed southwards, thought to be due to the basement of the Eastern Alps pushing in under the Pennine folds.

The eastern Pennides can be readily studied in the Chur district and along the routes leading to St. Moritz and Pontresina. The Simplon Pass provides easy access, and the cores of the Simplon-Ticino Nappes can be particularly well seen in the spectacular Gondo Gorge leading down to Italy. Zermatt is a good centre for examining the Pennide structures and the schistes lustrés with their ophiolite content, and the valley leading up to Arolla is a means of access to the Dent Blanche Nappe and Mont Dolin. Both the core and envelope of the Grand St. Bernard Nappe are well exposed along the road over the pass (2469 m) between Switzerland and Italy, which gives it its name.

7.8 The Northern Franco-Italian Alps - The Helvetides or Subalpine Chains
     (Fig. 7.10)

The Helvetic Nappes (7.5) can be traced south-westwards for some distance into France, where they are gradually replaced by parautochthonous and autochthonous folds of a relatively simple character usually overturned towards the N.W. and frequently broken by strike-faults. The succession is mainly Triassic to Oligocene with Miocene involved on the outer margin. These Subalpine Chains include the Haut-Griffe, Bornes and Bauges Massifs. Cliffs and jagged ridges are often made of Lower Cretaceous limestone, such as those of the Aravis. Some of the folds are characterised by décollement at the level of the Trias, Lias and Oxfordian. Towards the S., for instance E. of Chambéry, the structures are generally autochthonous and open.

The Thonon Syncline, in the Bornes, contains the Sulens and Annes Klippes, resting on Eocene or Cretaceous. Above a basal thrust a lower nappe of Ultrahelvetic facies consists of a Mesozoic succession followed by Eocene flysch. This is overridden by a Jurassic/Neocomian limestone/marly shale succession correlated with the sub-Briançonnais Zone and also with the Plastic Median Nappe of the Pre-Alps (7.4).

7.9 The Northern Franco-Italian Alps (Fig. 7.10) - The Hercynian or
     External Massifs and the Pennides

The north-easterly oriented Aiguilles Bouges and Mont Blanc Massifs form very high ground on either side of the Chamonix Valley, eroded along compressed Mesozoic, interpreted as the roots of Helvetic Nappes. The Massifs consist of metasediments, including the mica-schist which underlies the Mont Blanc summit ice-cap, probably largely of palaeozoic age and metamorphosed during the Hercynian Orogeny. These are invaded by granites (sometimes gneissose like the "protogene" of Mont Blanc), dated between 320 and 270 m.y. The crystalline rocks antedate Upper Westphalian sediments. The metasediments, however, include pre-Cambrian elements (Série de Fully in

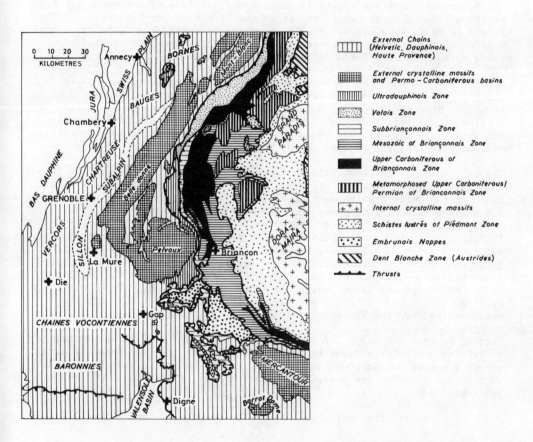

Fig. 7.10 STRUCTURE OF THE NORTHERN AND CENTRAL FRANCO-ITALIAN ALPS

the Aiguilles Rouges) which have been dated at between 770 and 730 m.y.

The Massifs are cut into wedges by thrusts inclined to the S.E. and often marked by mylonite. Several such zones were cut in the 11.6 km Mont Blanc road-tunnel which, with a rock cover of up to 2480 m, is the world's deepest.

These Massifs can best be studied from the Chamonix Valley where téléferiques give access to the higher outcrops.

The Belledonne Massif consists largely of mica-schists of Hercynian age (340 to 280 m.y.). Towards the S. end it is overlain by Upper Westphalian. The Massif is split by a narrow syncline of Mesozoic considered to be the continuation of the Chamonix Zone (see above). The Massif can be studied by going up the Isère or Doron valleys from Albertville.

About two miles down the Isère valley from Moutiers the External Massif of the Belledonne is in contact with the outermost unit of the Internal Alps or Pennides, here the Nappe des Brèches de Tarentaise. This nappe consists of a calcareous detrital series with conglomerates and breccias (erroneously referred to as flysch) in a recumbent syncline in unconformable but inverted contact with Lias and Palaeozoic cut by thrusts into slices.

The nappe belongs to the Valaisian Zone which fades out not far to the S. so that the Pennides, although structurally analogous to and continuous with those of the Swiss-Italian sector, consist largely in the French Alps of the Sub-Briançonnais, Briançonnais and Piémont Zones (see also 7.12). The Briançonnais Zone is characterised by major outcrops of the Palaeozoic, including unaltered or slightly altered Upper Carboniferous (Zone Houiller) externally and metasediments internally. In the present sector the Sub-Briançonnais Zone overrides the Valais Zone along a thrust and is separated from the Briançonnais Zone (consisting of several nappes) by the highly tectonised Nappe des Gypes, which has acted as a lubricant. The Zone Houiller flanks much of the Isère Valley upstream from Moutiers and the metamorphics form the Vanoise Massif.

Further E. come the Nappes of the Piémont Zone, characterised by schistes lustrés and ophiolites (seen in the Col de l'Iseran (2770 m)/Bonneval area) and by a gneiss basement making the Grand Paradis Massif. Traverses of this part of the Pennides are described by Ager et al., (1963, 505-510) and by Debelmas (1970, 51-100).

7.10 The Central Franco-Italian Alps - The Dauphinois Zone (Figs. 7.10, 7.11 and 7.12)

South of Chambéry, the Dauphinois and Provencal Zones, and a few Hercynian Massifs with their envelopes, make up the External Chains. Space does not permit a description of the complex Mesozoic and Tertiary facies-belts, but mention should be made of the Vocontian Trench, containing a thick marly Mesozoic succession, extending westwards from Gap. Its trend is an early manifestation of E.-W. folding which was an important phase in the southern Dauphiné and in Provence.

From Chambéry to Die the Chartreuse and Vercors Massifs (Arnaud, 1966), framed by Lower Cretaceous limestone cliffs, are characterised by N.N.E. folds with overriding along reversed faults towards the W.N.W. - similar to

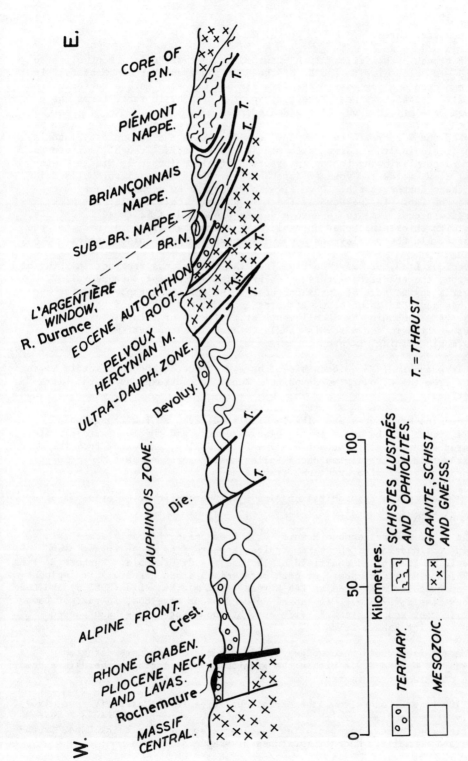

Fig. 7.11 SECTION OF CENTRAL FRANCO-ITALIAN ALPS FROM RHÔNE GRABEN TO PIÉMONT

the Jura structures.

A deep trench - the Sillon Alpin - extending southwards from Montmélian to beyond Grenoble, lies to the E. of the Chartreuse and Vercors Massifs. It has been eroded along soft Upper Jurassic but may also mark a tectonic line as small thrusts have been detected, and the trench, in fact, forms the western edge of the cover of the Belledonne.

South of the Vercors in the Devoluy, the Diois and the Baronnies (Flandrin, 1966) (including the sediments formed in the Vocontian Trench), structures are more complex owing to the interaction of older (probably Middle Cretaceous) E.-W. folds and younger (probably very late Miocene) northerly folds like those further north. Dome-like folds have been developed near Die, Veynes and Gap. In the Devoluy there is a large-scale overlap of thick Upper Cretaceous limestones onto older Mesozoic and it is these limestones, along with thickening Upper Jurassic, that form the cliffs in the S. of the Dauphiné. In the Devoluy/Gap area Oligocene occurs in folded basins.

The Ventoux Line, of overthrusting from the S., extends from not far S.W. of Sisteron, the traditional gateway of Provence across the Durance Valley, westwards to the edge of the Rhône basin. Near Digne a nappe of Trias and Lias has advanced about 6 km westwards over Miocene. Many of the Dauphiné folds are accompanied by décollements at the level of Triassic evaporites and Upper Liassic and Oxfordian shales. Above the two last-mentioned disharmonic folding may occur.

The Ultradauphinois Zone consists of narrow parautochthonous Mesozoic slices thrust from the E. over the Dauphinois Zone; it corresponds to the Ultrahelvetides.

The Dauphiné structures are clearly displayed, even at low topographic levels, owing to the more limited glacial drift, and the more arid climate, compared with the Alps further N. The Drôme Valley, eastwards from the most westerly fold at Crête, the Buchs Valley near Veynes and part of the Upper Durance Valley provide excellent traverses.

7.11 The Central Franco-Italian Alps - The Hercynian or External Massifs (Figs. 7.10, 7.11 and 7.12)

To the S. of the Belledonne Massif (7.9) Stephanian, probably above the hidden continuation of the metamorphics, appears from under Triassic and Jurassic in the worked coalfield of la Mure, S. of Grenoble. Further E. lies the Pelvoux Massif composed of both metasediments and granite; it includes such spectacular mountains as Les Ecrins (4102 m), La Meije (3983 m) and the Pelvoux itself (3932 m). It has an autochthonous envelope, in part of Lower Jurassic age, and on its S.E. side autochthonous Eocene has been pushed against it into sharp folds.

The Massif is divided into wedges by shear zones and thrusts directed towards the W. and N.W. Within the granite E. of Ailefroide there is a deep wedge of Jurassic sediments.

Some 100 km S.E. of Pelvoux the Mercantour (Argentera) Massif (Vernet, 1967), mainly of metasediments with a core of granite, is partly surrounded by red, clastic, continental Permian. The envelope also includes thick Upper Cretaceous and nummulitic Eocene limestones of simple structure.

Fig. 7.12 A. Tectonic wedge of Jurassic sediments (bedded centre) in Pelvoux Hercynian Granite. East of Ailefroide, Hautes-Alpes, France.

B. Fold of Lower Cretaceous limestone, Sisteron, Durance Valley, French Alps.

7.12 The Central Franco-Italian Alps - The Pennides (Figs. 7.10, 7.11, 7.13 and 7.14)

The Briançonnais Platform dominates this sector of the French Alps. The Platform consists mainly of Upper Carboniferous sediments and metasediments and separates the Valais Zone of deep sedimentation to the W. from the similar Piémont Zone to the E. The two zones are not, however, symmetrically preserved about the Platform, for the Valais Zone fades out in the northern French Alps (7.9) and the Piémont Zone becomes very wide in the present sector. The cover of the Briançonnais Zone is characterised by thick Triassic dolomite, followed by a Mesozoic succession, with breaks (e.g. the Lias is missing), which reveal the unstable nature of the Platform which, in fact, emerged at times from the sea.

The sub-Briançonnais Zone (Barbier and Debelmas, 1966) contains deeper water sediments deposited along the W. margin of the Platform. Lower and Middle Jurassic limestones and shales are important, followed by a succession up to the Eocene.

Tectonically, the sub-Briançonnais forms two or three narrow sheets. One of these, mainly of steeply dipping Middle Jurassic limestone, forms the spectacular Croix des Têtes, N. of the Arc Valley, near St.-Michel. Above the opposite side of the valley the Jurassic Breches du Télégraphe probably fell from the W. edge of the Platform. In this area the Nappe des Gypes (7.10) intervenes between the Croix des Têtes sheet and the Coal Measures of the Briançonnais Zone.

Much of the cover of the Briançonnais Zone has undergone décollement and has been folded on itself or superimposed as nappes (given a number of local names), which have moved relatively short distances. These structures were later refolded into a fan, on the eastern side of which they are overturned.

In the Window of l'Argentière, in the Upper Durance Valley, a complex culmination brings up Upper Jurassic to Upper Cretaceous limestones, shales, etc., of the sub-Briançonnais Zone within a frame of Triassic dolomites and quartzites, and Coal Measures, of the Briançonnais Zone. Although the structure may be only a "demi-fenêtre", (there is some doubt whether the Triassic dolomites on the W. completely close round the "fenêtre"), it shows that the Briançonnais Nappes have been pushed over the sub-Briançonnais by at least 10 km.

Minor structure in the l'Argentière district can be related to a first phase of flat folding, a second phase of nappe formation and a late phase of nearly upright folding (Fig. 7.13).

Further S., in a pitch depression, the Embrunais Nappes, of which the Flysch à Helminthoides Nappe consisting of Upper Cretaceous calc-sandstone and shale is the most important, overlap the Briançonnais and sub-Briançonnais Zones. These are believed to have moved W. from the Piémont Zone before the metamorphism to schistes lustrés. The nappes of the Piémont Zone consist largely of schistes lustrés with abundant ophiolites. The original tectonic superposition of the Piémont Zone on the Briançonnais is sometimes reversed by the later folding which formed the fan (see above). The crystalline basement of the Piémont outcrops S.W. of Turin, in the Dora Maria.

Further S., the Pennides become narrow between the overlapping molasse of

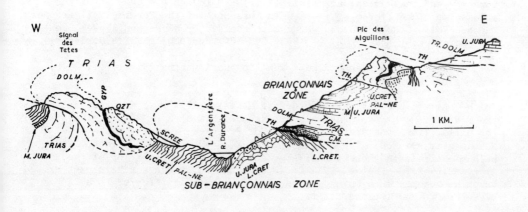

Fig. 7.13 SECTION OF WINDOW OF L'ARGENTIÈRE
(after Debelmas 1970)

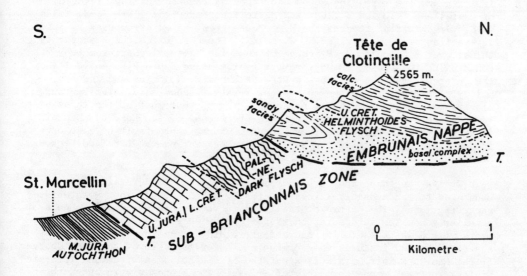

Fig. 7.14 SECTION ON NORTH SLOPES OF DURANCE VALLEY AT
ST. MARCELLIN
(after Debelmas 1970)

Lombardy and the Mercantour Massif (7.11). Owing to the overlap of the molasse no "Zone of Roots" is seen in this part of the Pennides. The Upper Durance Valley provides ready access to the sector described above; routes are suggested by Ager et al., (1963, 502-504) and Debelmas (1970, 54-181).

7.13   The Southern French Alps - The Provencal Chains (Figs. 7.1, 7.10, 7.15 and 7.16)

In the eastern part of this sector the Maritime Alps rise spectacularly from the Mediterranean coast and inland reach heights of over 3000 m. Westwards the hilly, limestone scenery, typical of Provence, extends to the Lower Rhône, near Avignon. The Provencal Chains (Goguel, 1953; Cornet, 1965) belong entirely to the External Alps and form the continuation of the sub-Alpine/Dauphinois Chains. The E.-W. trend noted in the southern part of the Dauphiné becomes much more dominant and is sometimes referred to as a Pyrenean-Provencal trend.

In the Maritime Alps the folds are disposed in the two major arcs of Nice and Castellane (Fig. 7.1). There is evidence for Upper Cretaceous and Upper Oligocene movements, but the main fold phase was late Miocene. This resulted in overfolding, generally towards the S., accompanied by over-thrusting in the same direction and the formation of nappes. Décollement is common on the Keuper. A marked tensional phase resulted in roughly N.-S. faulting and the formation of graben, such as that containing Oligocene S. of Castellane. To the W., between Castellane and Digne, late N.-S. folds with over-thrusting towards the W. produce complicated structures by their interaction with the earlier E.-W. folds. South of Digne the Mesozoic is thrust over Miocene/Lower Pliocene of the wide Valensole basin, which contains up to 1000 m of conglomerate representing a Mio-Pliocene delta of the Durance. Between this basin and the Rhône E.-W. folds and thrusts, mainly affecting Cretaceous to Miocene strata, predominate. The Ventoux line is marked by over-thrusting towards the N., but most of the thrusts are towards the S., for example, the Lower Cretaceous of the Luberon Mountains is thrust in this direction over Miocene. North-south faulting also occurs, for example, that forming the Sault graben separating the Lure and Ventoux Mountains.

Near Grasse the topography drops dramatically from the Jurassic limestone mountains to a dominantly Triassic terrain, which extends south-westwards past Draguignan. The nature of the Triassic strata has given this zone a peculiarly complex structure, characterised by sheets separated by décollements within which there is independent folding. The lowest of these sheets or "sole" includes the Permian and Lower Triassic of the N.-W. border of the Maures Massif. This is separated by a décollement from the Triassic "infrastructure", made up mainly of Triassic limestones within which there are two surfaces of décollement at the Lower and Upper Muschelkalk respectively. Interstratified with the Muschelkalk at Rougiers there is a basalt lava flow. The Jurassic "superstructure", occurring mainly in north-westerly synclines, is likewise separated by décollement from the Triassic.

Further W., in Low Provence (Fig. 7.15), dominantly E.-W. folds and thrusts, involving successions from the Lower Jurassic to the Miocene, prevail. Considerable thicknesses of Tertiary, ranging up to the Miocene, occur in synclines such as those of Aix and Marseilles. North of Aix there is a basalt lava flow of Upper Miocene age. Other small outcrops of Tertiary calc-alkaline volcanics, including basalts, andesites and trachybasalts

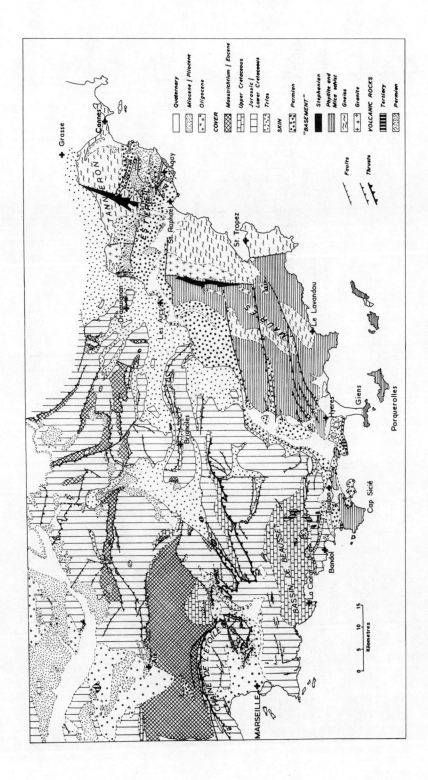

Fig. 7.15 STRUCTURE OF SOUTHERN FRENCH ALPS (PROVENÇAL CHAINS)
(after Debelmas 1974)

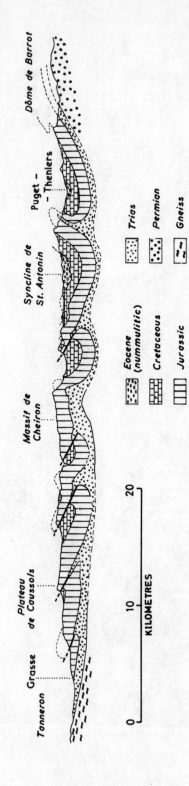

Fig. 7.16 SECTION OF SOUTHERN FRENCH ALPS FROM TANNERON TO DÔME DE BARROT (after Debelmas 1974)

occur in the Provencal coastal region. These are part of the "Outer Alpine" younger volcanics but can also be regarded as part of the Mediterranean Belt. Further S.E., the Le Beausset Basin contains an especially thick load of Upper Cretaceous sediments. North of Bandol Trias and Lias have been thrust northwards over the S. limb of this syncline.

In the southern part of Provence, including Low Provence just described, an E.-W. Provencal folding becomes more dominant and the main movements seem to have been Upper Eocene, possibly extending into the Lower Oligocene. The late Miocene movements did, however, have some influence, including some further folding and thrusting, and also fairly large-scale normal faulting. The overthrusting in southern Provence does not seem in general to exceed 1 km, and in this respect, as well as in the earlier date of the main movements, the southern Provencal Chains contrast with those further N., where the tectonics are more closely related to those of the sub-Alpine Chains generally.

7.14 The Southern French Alps - The Hercynian or External Massifs (Figs. 7.1 and 7.15)

The Maures Massif consists of a wide variety of metasediments, mainly pelites, with amphibolite, serpentinite, gneiss, migmatite and intrusive granite, of pre-Stephanian age. A number of groups have been recognised, including "Lower Mica-Schists" which contain garnet, staurolite and kyanite.

Apart from two N.-S. grabens of Stephanian, the cover of both the Maures and the Esterel Massifs consists of thick Permian. Red conglomerates and sandstones form the spectacular Roquebrune W. of Fréjus. The formation also contains abundant volcanics, including porphyry and felsite lavas, and thick, red, massive rhyolitic ignimbrites, which add to the beauty of the Esterel mountains and coast between St. Raphaël and Cannes. Further N. acid gneiss of the Tanneron Massif is well exposed in the hills W. of Cannes (Fig. 7.15). The Massifs are cut by numerous faults, mainly of N.-S. or E.-W. trends.

Smaller outcrops of Palaeozoic rocks also show from beneath a Permian cover further S.W. between Hyères and Bandol. These include sediments at Giens and on the Porquerolles, which contains Silurian graptolites, and phyllites at Cap Sicié, considered to be metamorphosed Silurian.

The Mediterranean coast provides good exposures, and other ready means of access include the Var Valley leading inland from Nice, and the Route Napoleon from Grasse through Castellane to Digne. Excursions are described by Middlemiss et al., (1970) and by Gouvernet et al., (1971).

The Pyrenees (France and Spain) (7.15 - 7.19)

7.15 Introduction (Figs. 7.17 and 7.18)

The Pyrenees form a 400 km long, narrow (40-80 km), mountain barrier between France and Spain which, like the Alps, has had a major influence on European history. The highest point, the Pico d'Aneto, reaches 3404 m, and several other peaks over 3000 m. The glaciers are smaller than in the Alps and the Pleistocene glaciations did not reach nearly as far down the valleys. As a result of this limited glacial erosion and the heavy tree cover of parts of the range, exposures are, on the whole, less spectacular than in the Alps.

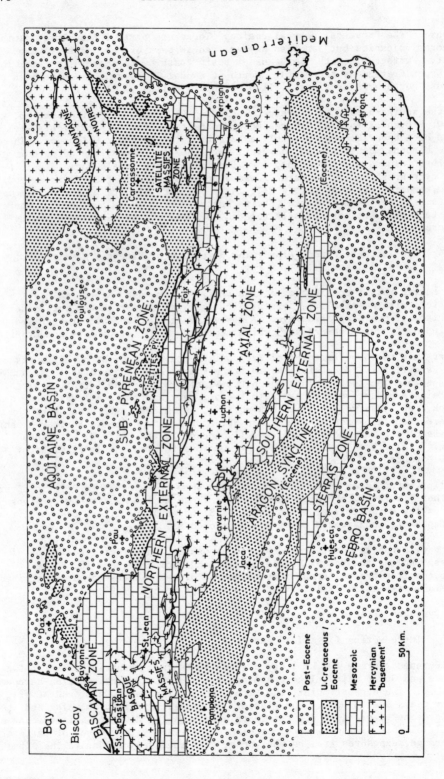

Fig. 7.17 STRUCTURE OF THE PYRENEES

A

B

Fig. 7.18 A. Recumbent fold in Upper Cretaceous limestone.
Coast between Biarritz and St.-Jean-de-Luz, France
B. Mica-schists in core of Betic Zone
North face of Mulhacen (3481 m, the highest point
in Spain), Sierra Nevada.

except in the dry eastern Spanish Pyrenees.

All major formations from late Precambrian to Pliocene are believed to be present in the Pyrenees. Geographically the Chain ends at the Atlantic, but Pyrenean folds extend along the S. coast of the Bay of Biscay (7.21).

The main structural units are:-

        Northern External Zone (France) (7.16)
        Satellite Massifs Zone (France) (7.16)
        Axial Zone (France and Spain) (7.17)
        Southern External (Nogueras) Zone and
        Sub-Pyrenaic Sierras Zone (Spain) (7.18)
        The Biscayan (Vascogothian) Zone (Spain) (7.19)

The principal structure-building events were:-

    HERCYNIAN   Early Westphalian (Sudetic)

                      (Lower Cretaceous - mainly tilting and uplift
    ALPINE   (Late Eocene
                      (Early Pliocene - mainly uplift

As Caledonian folding on any scale and Caledonian metamorphism are absent, Hercynian metamorphism is clearly identifiable. Moreover, Alpine metamorphism is weak, probably owing to much shallower burial of the Hercynian structures.

It has been suggested (e.g. De Sitter, 1964) that the Pyrenees were formed by an Alpine rejuvenation of an E.-W. Hercynian Chain. However, details of the Hercynian structures have shown trends quite distinct from the Alpine directions; furthermore, the Lower Palaeozoic of the Pyrenees shows considerable similarities with that of the Montagne Noire, reappearing from under the Tertiary some 50 km to the N. of the E. end of the Pyrenees. Neither the Cretaceous nor the Eocene show very marked differences on opposite flanks of the Pyrenees, nor any indication of wedging out, an observation borne out along the westward plunging nose of the Axial Zone. Other authors, therefore, (e.g. Rutten, 1969, 348) hold that the Pyrenees form a truly Alpine Chain with parts of a much more extensive Hercynian basement brought up in the Axial Zone.

The coastal routes round the ends of the Pyrenees and the main routes through the Chain, by St. Jean-Pied-de-Port, by the Somport Pass, and by Puigcerda, all provide good cross-sections. Details regarding the French Eastern Pyrenees are given by the Geological Society of France (1958), and excursion itineraries for parts of the Spanish Pyrenees by Rios and Hancock (1961).

7.16   The Pyrenees - Northern External and Satellite Massifs Zones
        (Figs. 7.18 and 7.19)

The Northern External Zone consists essentially of a succession from Jurassic to Eocene disposed in folds overturned and overthrust towards the N.; generally it is considered that the translation along the thrust does not exceed the order of 1 km. Smaller, often acute, folds are superimposed on the main folds. The northern front of the Zone often rises spectacularly

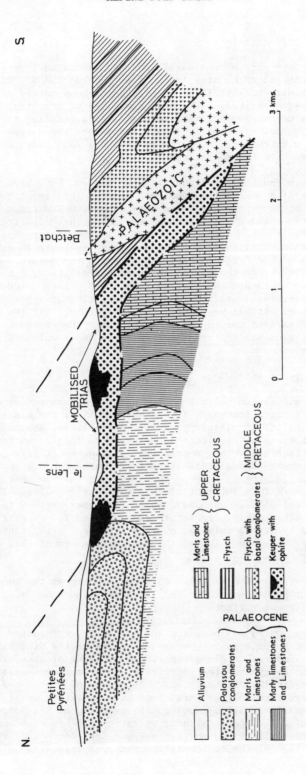

Fig. 7.19 SECTION OF NORTH PYRENEAN THRUST FRONT

above the plains of Aquitaine.

North of the eastern part of the Axial Zone and separated from it by a major fracture zone, thought to be steeply inclined towards the S., known as the North Pyrenean Fault, the Satellite Massifs Zone shows marked differences from the zone just described. It contains massifs of granites, gneiss, schists and less metamorphosed Palaeozoic sediments, similar to those of the Axial Zone, surrounded by strongly folded Mesozoic; the facies of the latter contrast with that of the Northern External Zone, particularly in the presence of very thick Lower Cretaceous.

The North Pyrenean Fault may well be transcurrent and have brought into contact two crustal blocks which were once considerably separated; the fault, in fact, might be sinistral and have brought westwards a block from the eastern continuation of the Pyrenees now under the Mediterranean. The North Pyrenean Fault may continue westwards. In the western Pyrenees, S. of the Aquitaine Basin, drilling has shown a complex fracture zone, probably the continuation of the Pyrenean Fault, over which the Mesozoic has been draped (Henry, 1967). To the N., as shown by the drilling, a series of open folds occurs with northward overturning. To the S. of the hidden fault overfolds and overthrusts are towards the S. and cleavage, absent to the N., has been developed. It has been suggested, therefore, that the North Pyrenean Fault separates two major blocks of the Pyrenees and is perhaps a more significant tectonic feature than the Axial Zone, which is due to late uplift (see below).

7.17 The Pyrenees - Axial Zone (Fig. 7.17)

The Pre-Permian succession falls into two parts which have tectonic significance. The lower part, downwards from the Caradocian, the oldest formation with fossils, consists of clastic psammites and pelites, several kilometres thick, seen, for example, near Luchon, believed to range down to the Upper Precambrian. Tectonically this succession is disposed in large, simple, dome-like structures; among these syntectonic magmatism and metamorphism have produced mantled gneiss domes. The Silurian, for the most part, consists of not more than 100 m of black, graptolitic shales. This has acted as a level of décollement above which the upper part of the succession, consisting of shales, marls, limestones and sandstones, not over 1 km thick, is characterised by strong folding, overthrusting and minor nappes. Particularly striking folds occur in the thick Dinantian limestones, for example N. of the Somport Pass.

In the eastern Pyrenees the tectonic distinction between the two successions is not as marked, probably because the Silurian is more sandy and has not acted as a décollement surface.

Hercynian folding, which may have started in the Upper Devonian, reached its peak in the Lower Westphalian. The Axial Zone is at its widest in the E. There is a pitch depression along the frontier E.S.E. of St. Jean-Pied-de-Port and then the Hercynian structures rise again to form a massif S. of San Sebastian, which finally pitches beneath Jurassic and Cretaceous at Tolosa. The Hercynian structures of the Axial Zone have a limited cover of Permo-Triassic continental red beds, exposed, for example, at the Somport Pass; the Triassic includes evaporites, some volcanics and basic intrusions. Marine Mesozoic deposits range up to the Upper Cretaceous and are significant in showing that Alpine sedimentation extended across the old Hercynian

Chain.

The Cerdana Basin near Puigcerda contains a Miocene/Pontian succession. The presence of lignites, evidently deposited in marshy lowlands, at 1700 m, is part of the evidence for the post-Pontian uplift of the Axial Zone.

Two phases of syntectonic magmatism and metamorphism have been recognised. The earlier phase formed metamorphic domes containing gneisses and micaschists and did not cut higher than the top of the Ordovician. The later phase, which penetrates locally into the Devonian, was one of migmatisation and granitisation with the formation of leucogranites. Some of the gneisses form spectacular scenery such as the Canigou, 2785 m, in the French Eastern Pyrenees. In places, however, even the older Palaeozoic strata have escaped metamorphism, as seen, for example, on the coast at the E. end of the Pyrenees at Cerbère, or have undergone phyllite grade metamorphism as at the Pas-de-Roland, S. of Bayonne.

Late, cross-cutting granites penetrated into the Carboniferous. These form some of the most spectacular scenery of the Pyrenees, such as that of the Andorra Massif which continues eastwards to form the striking scenery near Nuria, and of the Maladetta Massif in the central Pyrenees, which includes the highest peak of the Chain.

7.18 The Pyrenees - Southern External (Nogueras) and Sub-Pyrenaic Sierras Zones (Figs. 7.17 and 7.20)

The northern border of the Southern External Zone is formed by the steep, northerly-inclined South Pyrenean Fault. This may have an important horizontal component as suggested by the way some granites of the Axial Zone are cut off. The Zone is characterised by sharp folds and overthrusts towards the S., on which the overriding is of limited extent. Some of the structures have been attributed to gravity tectonics following the late uprise of the Axial Zone. The strata affected are mainly Upper Cretaceous and Oligocene; much of the Upper Cretaceous rests directly on Carboniferous, for example at Canfranc, and the Jurassic is only sparingly developed.

The border between the Southern External Zone and the Axial Zone is famous for one major nappe, that of Gavernie. Upper Cretaceous limestone, forming the spectacular Cirque de Gavernie on the frontier, extends northwards down the R. Pau and up its tributary, the Heas, resting on Lower Palaeozoic, probably Ordovician. The Upper Cretaceous has been overthrust by slightly metamorphosed Palaeozoic metasediments, with fossils ranging from Silurian to Carboniferous. The two valleys form a demi-fenêtre which shows that the overthrusting must have been at least 10 km.

A zone of overthrusting towards the S. forms the southern boundary of the Southern External Chains, between which and the Sierras, W. of Ainsa, there is a major synclinorium with sharp folds in thick Eocene and Oligocene marly shales, marly sandstones and hard sandstones. This forms the relatively low ground past Jaca to Pamplona, diversified by the Peña de Oreol, of coarse Eocene molasse. The Sierras consist of fairly open folds, slightly overturned towards the S. During the Mesozoic the Sierras Zone was extremely mobile as shown by the great variations in thickness and in continuity of the succession.

In the Montsech area the Upper Cretaceous is very thick and rests on Lower

184 STRUCTURE OF WESTERN EUROPE

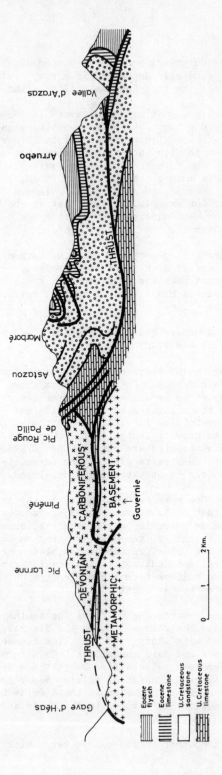

Fig. 7.20 SECTION OF GAVERNIE THRUST

Jurassic. Further W., where the Rio Gallego bends southwards at Ste. Maria La Pĕna, close to the S. front of the Sierras, Upper Cretaceous limestone, dipping steeply N., rests directly on red Triassic sandstones and marls. The latter formation in the Sierras contains thick salt/gypsum beds which have complicated the tectonics by forming numerous diapirs, notably in the Montsech area. A thrust over the Ebro Basin (8.3) forms the southern front of the Sierras.

7.19   The Pyrenees - The Biscayan Zone

The Southern External Zone curves round the W. end of the Axial Zone and continues along the S. coast of the Bay of Biscay to beyond Santander. Sharp folds broken by steep strike faults affecting mainly Lower Cretaceous strata are seen along the coast. Downfolds W. of Bilbao and at San Sebastian bring down Upper Cretaceous overlain by Eocene sandstone. The Lower Cretaceous becomes very thick S. of Tolosa, in the Sierra de Aralar, and overlies dark Jurassic shaley limestones. Further W., both Lower and Upper Cretaceous are very thick in the Cantabrian Trough, N. of Vitoria. The underlying Keuper has given rise to salt domes which penetrate up to the Oligocene. This is an area where oil exploration has been undertaken. The location of the diapirs is independent of the fold tectonics, but there seems to be some tectonic control as many of the salt domes originate along the line of very rapid thinning of Lower Cretaceous sediments to the S. and E., which is thought to correspond to a big fault zone in the Palaeozoic basement underneath (Rios, 1961, 371).

7.20   The Catalonian and Iberian Chains (Spain) (Figs. 5.8 and 7.21)

After the Hercynian orogeny, the Iberian Peninsula, N. of the Guadalquiver Basin, became divided into a western and an eastern region with very different geological histories. The westerly region, possibly owing to the presence of a higher proportion of granites and metamorphics, remained relatively rigid; in the more easterly, however, two geosynclines developed as a frame (along with the Pyrenees) of the large triangular Ebro Block, later to become the Ebro Basin.

The marine and continental sediments deposited in these geosynclines were folded to form the north-easterly Catalonian Chains, S.E. of the Ebro Basin, and the north-westerly Iberian Chains, S.W. of the Basin, both rising to over 2000 m. Where the two chains meet, S.W. of Tarragona, structures show a curvature from one trend to the other.

The Palaeozoic rocks of the chains may range from Ordovician to Upper Carboniferous and in the Catalonian Chains there are post-tectonic granitic rocks. The younger rocks range from Triassic to Eocene in the fold-belts, with younger Tertiary in late grabens. The Trias includes the Muschelkalk with several gypsum horizons. The higher Mesozoic and Eocene are mainly epicontinental carbonate rocks with some strata of Wealden facies.

Structure-building phases are:-

    HERCYNIAN   Early Westphalian (Sudetic)

                      (Late Eocene (N.E. part of Catalonian Chains)
    ALPINE   (Late Oligocene
                      (Plio-Pleistocene - mainly uplift and faulting

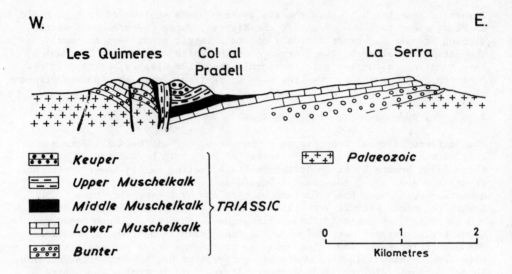

Fig. 7.21 SECTION OF PART OF CATALONIAN CHAINS NEAR TARRAGONA, SPAIN.

The N.E. and central parts of the Catalonian Chains are divided by the Grenollers Graben into the Coastal Chains, extending past Barcelona, and the Pre-Littoral Chains stretching S.W. from Gerona. There is considerable exposure of "basement", including both Palaeozoic and granite, on which the Bunter sandstone rests almost horizontally. Décollement occurs at several evaporite horizons in the Muschelkalk, above which are epidermal folds of Jura type. Similar folds occur in the S.W. part of the Chains where, however, there is a greater preservation of Mesozoic, including limestones which form jagged, bare scenery. Fold-structures do not individually persist far along the strike due to segmentation into blocks by cross-faults, some of which can be seen to cut the basement, blocks of which can also be shown to be tilted.

Some Mio-Pliocene basalt flows occur near Gerona. These are the only Tertiary lavas in the Alpine Fold-Belts of Northern Spain and their position where the Pyrenean and Catalonian Chains meet near the Mediterranean is probably tectonically significant.

The Iberian Chains (Burkmann, 1962) are also split by a major graben, in their case the Calatayud-Teruel depression. At its N. end, around Calatayud, this is a simple north-easterly graben flanked by large, elongated blocks of Palaeozoic metasediments. Further S. it is broken up into a number of depressions. The graben is largely floored by Miocene sediments, containing thick, gypsiferous evaporite, well seen at Calatayud itself. Several other blocks of Palaeozoic, both metamorphic and non-metamorphic, occur in the Chain. Their relationship to the overlying Triassic shows that, as in the case of the Catalonian Chains, the folding is of the Jura type; overfolding and overthrusting are mainly towards the N.E., involving strata up to the Lower Oligocene. Along parts of its northern and north-eastern margin the Chain has been pushed over the Ebro Basin. There is the same fragmentation

of the structures, and twists of strike occur, for example near Soria, which
may be due to differential movements in the basement. Cretaceous limestones
form a large part of the Chains in this area and the folds have been strongly
denuded giving rise to plateau-like scenery.

The Iberian Chain and the Catalonian Chains do not rise to the heights of the
Pyrenees, the highest point being just over 2000 m, and they are traversed by
a number of routes which provide ready access. There are long coastal
exposures of parts of the Catalonian Chains.

The Betic Cordillera (Spain) (7.21 - 7.23)

7.21   The Betic Cordillera - Introduction (Fig. 7.22)

The high Betic Cordillera are major features of the striking scenery of
Andalucia, the hot "North African", southern part of Spain, ruled by the
Moors until the fall of Granada in 1492. The Mulhacen, the crest of the
Sierra Nevada in the heart of the Cordillera, rises to 3481 m, the highest
point in the Iberian Peninsula.

The western part of the Cordillera drops steeply to the Guadalquiver Basin
(8.3); the eastern is in contact with the S. end of the meseta or Iberian
Platform and, near the Mediterranean, cuts sharply across the north-north-
westerly folds at the S. end of the Iberian Chains.

From N. to S. the Betic Chains are divided into the following units:-

> The Prebetic Zone   )
> The Subbetic Zone   )   (7.22)
> The Betic Zone (senso stricto) (7.23)
> The Gibraltar Complex or Ultrabetic Zone (7.23)

The oldest rocks of the Cordillera are the metasediments of the Betic Zone,
possibly of Palaeozoic age, although a Precambrian age has also been suggested. These consist of mica-schists, with garnets in places, quartz-mica-
schists and some quartzites. Fossiliferous Silurian to Carboniferous strata
also occur.

All the main sub-divisions of the Mesozoic and Tertiary are present, but in
the central part of the Cordillera there is no Mesozoic above the Triassic.
This formation shows two major facies which have tectonic significance. In
the outer ranges the "Germanic" facies is present, including the Muschelkalk,
with gypsum beds which have acted as a lubricant. In the inner ranges a
thick carbonate facies of Alpine-type occurs.

The younger Tertiary is confined to post-folding basins and grabens which
diversify all but the highest parts of the Chains. Thick gypsum beds occur
in the Miocene and are worked at several localities.

Molasse sandstones and conglomerates, derived from the earlier parts of the
Chains, occur in most of the basins and grabens. Some of them build conspicuous, isolated hills and mountains. Basic volcanics, probably Miocene, occur
on the Mediterranean coast E. of Almeria.

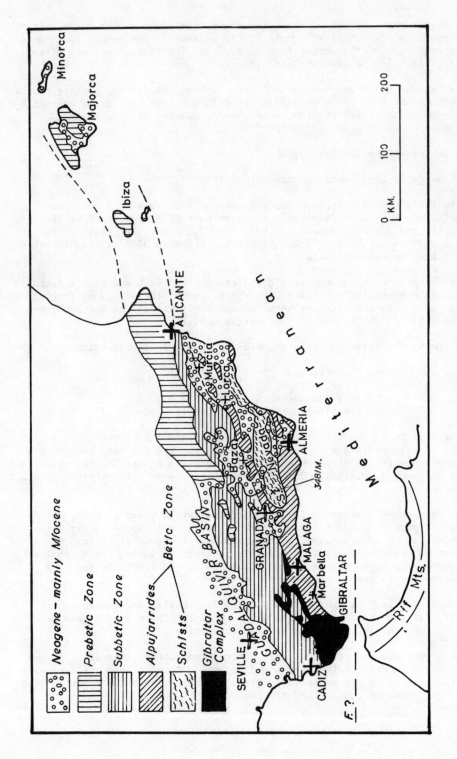

Fig. 7.22 STRUCTURE OF THE BETIC CORDILLERA, SPAIN

The following are the main structure-building events:-

    HERCYNIAN   Pre-Triassic

                          (Late Cretaceous/Early Eocene
    ALPINE   (Late Oligocene/Early Miocene
                          (Plio-Pleistocene - uplift and basin/graben formation.

Coastal exposures provide good sections of the southern margin of the Betic Cordillera and the lower parts are generally traversed by a number of transport routes. The spectacular Sierra Nevada road goes almost to the summit of Valeta, 3401 m, the second highest peak. Most of the topography of the Betic Cordillera is unique for a Western European, high mountain chain in the absence of the effects of Pleistocene glaciation. Glaciation did occur, however, on the N. face of the summit ridge, as corries and small moraines are present.

7.22   The Prebetic and Subbetic Zones (Fig. 7.22)

The Prebetic Zone is developed only in the north-eastern part of the Chain and is characterised by fairly simple folding on N.E. to N.N.E. axes. Along its N.W. margin overthrusting occurs in the same direction. The bare, Mesozoic limestones, which are a feature of the scenery of the Chain N. of Alicante, continue north-eastwards through Ibiza to the Balearic Islands, and in Majorca overthrusting towards the N.W. has also been recognised.

In the Subbetic Zone, which is thrust northwards over the Prebetic, the structures become more complicated but are still autochthonous or parautochthonous; they are of the epidermal type, riding on a décollement controlled by the gypsum horizons in the Triassic. The tectonic style is that of the Sub-Alpine Chains. Along the inner edge, a sub-zone, the Penibetic, is recognised by some authors.

7.23   The Betic Zone, the Gibraltar Complex and the Straits of Gibraltar
        (Fig. 7.22)

The lowest element in this Zone is the Sierra Nevada Betic forming the high central ridge, structurally a late dome of the Oligocene/Miocene nappe structures characteristic of the Betic Zone as a whole. Most of the Sierra Nevada consists of a lower series of schists up to garnet grade; there is an upper series of felspathic schists along with intercalations of marble and of gypsum, which is probably metamorphosed Triassic. Some authors have assigned the Sierra Nevada schists to a nappe system named the Nevado-Filabrides, but there is no evidence to suggest that the schists are not autochthonous (Fig. 7.18).

Surrounding the Sierra Nevada and overthrust from the S. are the Alpujarrides, in which there are probably several nappes. These seem to consist of two series, an older of mesozonal schists with some meta-limestones, and a younger, which is the more conspicuous, of thick Triassic dolomites which form spectacular scenery on the northern side of the range. Extensive masses of peridotite occur around Marbella. Their structural position is doubtful, but they are thought to be an element of the Alpujarrides, possibly derived from ocean floor crust.

Between Ronda and Malaga a tectonically separated sheet, sometimes called

the Rondaides, consists only of Upper Triassic and Lower Jurassic. The
thick dolomites in the Triassic suggest that it has been derived from the
Alpujarride domain.

The Malaga Betic is thrust over the Alpujarrides (Boulin, 1963) from the W.
and continues northwards round them, so that it also overrides the Subbetic.
Although there is a Mesozoic succession from the Lower Triassic to the Lower
Tertiary, this nappe is characterised by the predominance of Pre-Alpine
rocks. These consist of an older group of high-grade metamorphics, including gneisses and calsilicate rocks, and a younger, slightly metamorphic
series of clastic rocks with fossils ranging from Silurian to Carboniferous.

In the Campo de Gibraltar, Cretaceous/Lower Eocene flysch forms a nappe-
complex - the Gibraltar Complex - which is thought to be of Ultrabetic
origin. It has also been interpreted as a sheet which has moved into a
structural depression by gravitational gliding. The Rock itself, however,
is formed of an inverted Jurassic succession, mainly limestone, which may be
a klippe resting on the flysch (Bailey, 1953, 157).

Gibraltar is by far the nearest part of Europe to Africa and this raises the
problem of the continuity of geological structure. Many authors accept that
the geology of the Rif Mountains is sufficiently similar to that of the
Betic Cordillera for the two to be parts of the same arc, formed as late as
the Oligocene. This implies that the Straits of Gibraltar, considered to
hide a transform fault which continues past the Azores to the Mid-Atlantic
Ridge, are geologically young. Study of evaporites interbedded with deep-
water sediments, from Mediterranean deep-sea cores, has suggested that the
Straits of Gibraltar closed and opened at least eight times in the Upper
Miocene (Hsu, 1973). At the beginning of the Pliocene final opening of the
Straits followed rifting along the transform fracture.

REFERENCES

Ager, D.V. & Evamy, B.D. 1963. The Geology of the Southern French Jura.
P.G.A., 74, 325-355.
Ager, D.V. et al. 1963. Summer Field Meeting in the French Jura and Alps,
July 1963. P.G.A., 74, 483.
Arnaud, H. 1966. Contribution à l'étude géologique des plateaux du Vercors
meridional. Géol. Alpine, 42, 33-52.
Aubert, D. 1959. La décrochement de Pontarlier et l'orogénèse du Jura.
Mem. Soc. vaud Sci. Nat., 12, 93-152.
Bailey, E.B. 1935. Tectonic Essays: mainly Alpine. Oxford.
Bailey, E.B. 1953. Notes on Gibraltar and the Northern Rif. Q.J.G.S., 108,
157-175.
Barbier, R. & Debelmas, J. 1966. Reflexions et vues nouvelles sur la zone
subbriançonnaise au Nord du Pelvoux (Alpes occidentales). Géol. Alpine,
42, 97-107.
Boulin, J. 1963. Sur les Alpujarrides occidentales et leurs rapports avec
la nappe de Malaga (Andalasia occidentale). Bull. Soc. Géol. France,
7(4), 384-389.
Burkmann, R. 1962. Aperçu sur les chaînes iberiques du Nord de l'Espagne.
In: Delga (ed.), Livre à la Memoire du Professeur Paul Fallet. Soc. Géol.
France, Paris, 291-300.
Buxtorf, A. 1908. Geologische Beskreibing des Weissensteintunnels und
seiner Umgebung. Mem. Carte Geol. Suisse, 21.

Buxtorf, A. 1916. Progresen und Befunden beim Hauensteinbasis und Grenehen-
   bergtunnel und die Bederrtung der letsteren fur die Geologie des
   Juragebirges. Verhandl Naturforsch. Ges. Basel, 27, 184.
Campredon, R. & Boucarut, M. 1975. Alpes-Maritimes, Maures, Esterel.
   Guides Géol. Rég., Paris.
Cavet, P. 1959. Stratigraphic du Palézoique de la zone axial pyreneenee a
   l'est de l'Ariège. Bull. Soc. Géol. France, 6(8), 853-867.
Chauve, P. 1975. Jura. Guides Geol. Reg., Paris.
Cornet, C. 1965. Evolution tectonique et morphologique de la Provence,
   depuis l'Oligocene. Mem. Soc. Géol. France, N.S., 44, 2, 103, 1-252.
Debelmas, J. 1970. Alpes (Savoie et Dauphiné). Guides Géol. Reg., Paris.
Debelmas, J. & Lemoine, M. 1970. The Western Alps: Palaeogeography and
   Structure. Earth-Sci. Rev., 6, 221-256.
De Sitter, L.V. 1964. Structural Geology. 2nd ed. McGraw-Hill, New York.
Egeler, C.G. 1964. On the tectonics of the eastern Betic Cordillera (S.E.
   Spain). Geol. Rundschau, 53, 260-269.
Fallet, P. et al. 1961. Estudies sobre las seines Sierra Nevada y de la
   Cramada Mischungs Zone. Bol. Inst. Geol. Minera Espan., 71, 345-557.
Flandrin, J. 1966. Sur l'âge des principaux traits structuraux du Diois et
   des Baronnies. Bull. Soc. Géol. France, 7, 376-386.
Gignoux, M. & Moret, L. 1952. Géologie dauphinoise. Masson, Paris.
Goguel, J. 1953. Les Alpes de Provence. Hermann, Paris.
Gouvernet, C. et al. 1971. Provence. Guides Géol. Rég., Paris.
Guttard, G. 1959. La Structure du Massif du Canigou. Bull. Soc. Géol.
   France, 6(3), 907-924.
Gwinner, M.P. 1971. Geologie der Alpen. Stuttgart.
Henry, J. 1967. Le probleme des étages tectoniques dans les Pyrenees
   occidentales. Comparaison entre les occidents nord-Pyreneennes et
   l'occident de Saureau-Gourette. In: J.P. Schaer (ed.), Colloque Etages
   Tectoniques. Baconniere. Neuchatel. 253-267.
Hope Hado, N. 1947. Contribucie al Conocionente de la Morfoestructuracle
   las Catalonides. Conseja Super. Cient. Inst. Lueas; Mallada, Madrid.
Hsu, K.J. et al. 1973. Late Miocene dessication of the Mediterranean. Nat.,
   242, 240-244.
Jager, E. et al. 1967. Rb/Sr Alterbestinimengen an Glimmern der Zentral-
   aefen. Mater. Carte. Geol. Suisse, N.S., 134.
Kerchhove, C. 1969. La "Zone du Flysch" dans les nappes d'Embrunais-Ubaye
   (Alpes occidentales). Geol. Alpine, 45.
Lienhardt, G. 1962. Géologie die bassin houiller Stephanian du Jura et de
   ses morts-terraine. Mém. Bur. Rech. Géol. Minières 9.
Middlemiss, F.A. et al. 1970. Summer Field Meeting in Provence, 1967.
   P.G.A., 81, 363-396.
Ramsay, J.G. 1963. Stratigraphy, Structure and Metamorphism in the French
   Alps. P.G.A., 74, 357-391.
Rios, J.M. 1961. A Geológical Itinerary through the Spanish Pyrenees.
   P.G.A., 72, 359-371.
Rios, J.M. & Hancock, J.M. 1961. Summer Field Meeting in the Spanish
   Pyrenees, 1961. P.G.A., 72, 373-390.
Rosset, L. 1968. Pointe de vue nouveau sur la structure des klippes de
   Savoie. Geol. Alpine, 44, 333-338.
Rutten, M.G. 1969. The Geology of Western Europe. Elsevier, Amsterdam.
Soc. Geol. de France. 1958. Réunion Extraordinaire dans les Pyrenees
   Orientales. Bull. Géol. Soc. France, 6e ser., 8, 805-978.
Soc. Géol. de France. 1961. Séance sur les Cordilleras betiques. Bull. Soc.
   Géol. France, 7(2), 263-362.

Soc. Géol. Suisse. 1967. Guide géologique de la Suisse. Wepf & Co., Bâle.
Steck, A. 1968. Petrographische und tectonische Untersuchungen am Zentralen Aargranites der Westliehen Aarmassivs. Eclogae Geol. Helv., 61, 19-48.
Trumpy, R. 1955. Remarques sur la Correlation des Unités Penniques Externes entre la Savoie et le Valais et sur l'Origine des Nappes Préalpines. Bull. Géol. Soc. France, 5, 217-231.
Trumpy, R. 1963. Sur les racines des nappes helvétiques. In: Livre Fallot, 2, Géol. Soc. France, Paris, 419-428.
Trumpy, R. 1969. Die helvetischen Deeken der Ostschweiz. Eclogae Geol. Helv., 62, 105-142.
Vernet, J. 1967. Le massif de l'Argentera. Géol. Alpine, 43, 193-216.
Virgill, C. 1958. El Triassico de les Catalonides. Bol. Inst. Geol. Minera Espan., 69.
Wenk, E. 1956. Die Sepontinische Gneissregion und die jungen Granite der Valle della Mera. Eclogae Geol. Helv., 49, 251-265.

Chapter 8
MESOZOIC AND TERTIARY SEDIMENTARY BASINS AND OIL/GAS FIELDS

8.1 The Saône Graben and the Rhône Corridor (France) (Figs. 8.1 and 8.2)

From the S. end of the Vosges (5.5) to the mouth of the Rhône the Alpine Chains described in the last Chapter are bordered to the N.W. and W. by a tectonic depression which, for most of its length, separates the fold-mountains from Hercynian blocks (5.1 - 5.5). The most important elements are the Saône Graben, followed by the Saône from N.E. of Dijon to Lyon, and the Rhône Corridor (couloir rhôdanien) through which the Rhône flows from Lyon to the sea. These depressions form the major part of an easy route from Paris to the Mediterranean.

Different parts of the system are of different ages, and the Rhône Corridor did not come into existence, as such, until the Miocene. There are good exposures on parts of the step-faulted flanks and in the Rhône valley between Montélimar and Orange, but in general alluvium covers the floors.

Some excursions are described by Middlemiss and Moullade (1970).

Exposed formations are :-
               Pliocene
               Miocene
               Eocene
               Cretaceous
               Jurassic
               Triassic
               Hercynian granite and schist

The Hercynian rocks are seen in the tiny "massif" of Serre (5.4), S.E. of Dijon, and in small strips towards the W. side of the Saône Graben between Chalons and Macon.

Triassic outcrops in the Macon district include M. Triassic (Muschelkalk) arkoses and U. Triassic (Keuper) marls with salt and gypsum. In the Rhône Graben, likewise, Triassic deposition did not begin until the Muschelkalk. Outcrops along the S.W. trending border from Valence to Alès show arkoses and sandstones followed by carbonates and marls with salt and gypsum, above which comes a dominantly sandy Keuper succession.

Fully marine conditions were established in the Rhaetic or Lower Lias. Detailed study of Jurassic facies-variation has made it possible to work out a complex palaeogeographic history. Broadly, the Jurassic consists of marls and limestones, including bioclastic and reef facies. In the Saône Graben ironstones, formerly worked, occur in the U. Jurassic.

The Jurassic of the Saône Graben is continuous with that of the Paris Basin (8.6) through the Dijon "strait", where the Seuil de Bourgogne anticline marks the structural division between the two regions.

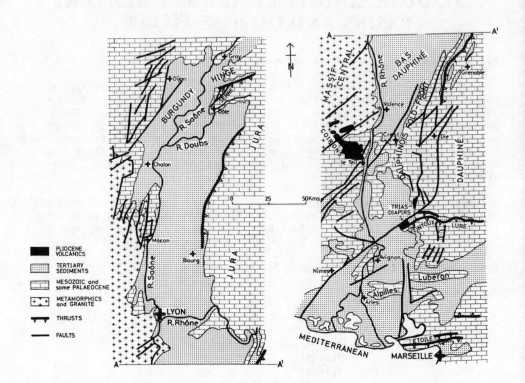

Fig. 8.1 STRUCTURE OF THE SAÔNE GRABEN AND OF THE RHÔNE CORRIDOR.
(after Debelmas and Demarcq 1974)

In the E. of the Rhône depression the Jurassic succession is similar to that of the Dauphiné Alps (7.10), with thick, dark Oxfordian marls. West of the river limestones predominate.

After emergence at the end of the Jurassic (Portlandian) a shallow sea spread over the area, leading to the deposition of marls and limestones in the Saône Graben and the northern part of the Rhône Graben. Further S., however, these give way to the hard, thick limestones of Urgonian facies, which make wide, conspicuous outcrops across the Lower Rhône valley. The junction between the two facies is not transverse to the depression but N.E. to N.N.E., indicating the early establishment of this Alpine trend.

Pre-Albian deformation led to uplift and erosion before the deposition of

Albian of greensand facies. In the N. the Upper Cretaceous is similar to
that of the Paris Basin. The extensive outcrops of Upper Cretaceous in the
Lower Rhône valley, however, consist of littoral or lagoonal deposits,
including sands, clays and lignites.

Uplift, mild folding and erosion preceeded the deposition of the Tertiary.
This starts with detrital beds, especially in the S., but much of the Eocene
and Lower Oligocene consists of lacustrine limestones. Upper Oligocene
deposition was disturbed by the main phases (see below) of the establishment
of the rift-valleys. In the Miocene, however, there was invasion by a
shallow sea in which mainly sandy sediments were deposited. Pliocene clays
and sands are also widespread.

Quaternary deposits are widespread in the Saône and Rhône valleys,
especially at the delta of the Rhône which includes the great plain of the
Camargue, only 4 m above sea-level. Here the Quaternary deposits cover very
thick Pliocene marine clays infilling pre-Pliocene valleys which extend to
1000 m below sea-level. Their formation is related to the Neogene history
of the Mediterranean (7.23).

Unlike the Rhine Graben, the Rhône depression has no volcanoes on its floor,
but the Pliocene volcanics (9.3) of the Plateau de Coirons make its W. flank.

Tectonic Evolution

Pre-Mesozoic structures are not seen in the Saône and Rhône depressions, but
the posthumous effect of Variscan lineaments is evident from the numerous
N.E. tectonic directions. During most of the Mesozoic Alpine movements were
confined to uplift or subsidence. Folding, however, occurred in Mid-
Cretaceous (Pre-Albian) and Late-Cretaceous times.

As a result of the powerful, late Eocene Provençal movements (7.13), the
floor and margins of the southern part of the Rhône depression are characterised by E.-W. folds.

Later (Mio-Pliocene) Dauphinois compression (7.10) formed N.E. to N.N.E.
folds N. of Valence.

The grabens were formed as such by faulting mainly at the end of the
Oligocene. The Saône Graben (or fosse bressan) complex in detail, with
subsidiary basins, is shaped by a N.-S. fault-system which swings into a
N.E. direction as the graben curves round towards the Burgundy Gate. To the
E. the Juras (7.2) have overridden the graben along a thrust of early
Pliocene age.

The subsidence of the northern part of the Rhône is due more to downwarping
than faulting, but S. of Valence the western margin has been depressed along
a number of N.E. faults, probably reactivated Hercynian fractures. The Bas-
Rhône fault, to the S.E. of the extensive Nimes fault, has a downthrow, as
proved by bores, of 5000 m (Debelmas and Demarcq, 1974, 522). The north-
easterly Alès graben contains a large Oligocene outcrop.

8.2   The Rhine Rift-Valleys (France and West Germany) (Figs. 8.3 and 9.6)

The Upper Rhine graben (Le Fosse Rhénan), which extend N.N.E. for 300 km
from Bâle to Mainz, with a width of 35 to 40 km, is the most striking

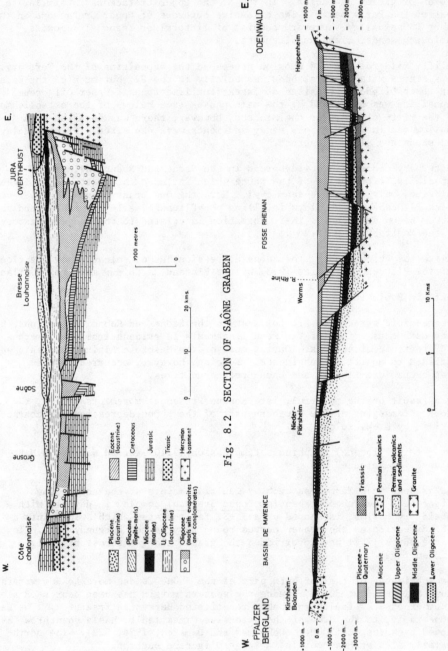

Fig. 8.2 SECTION OF SAÔNE GRABEN

Fig. 8.3 SECTION OF RHINE GRABEN

example of a rift-valley in Europe. Beyond Mainz the course of the Upper Rhine graben is continued by the Hessen Graben. To the N.N.W., however, a series of faults, cutting through a recently uplifted part of the Rhine Slate Mountains (5.6), ties up with the Lower Rhine Depression (la Rhénanie). This opens out north-westwards onto the Low Countries Plain (8.4).

The rift-valleys provide easy transport routes from Switzerland and Central Europe to the North Sea.

Apart from volcanoes, the floors of the rifts do not rise much above 250 m; on the flanks of the Upper Rhine graben the Vosges (5.5) and the Schwarzwald (5.5) reach nearly 1500 m. The rocks on the upfaulted flanks range from Precambrian to Jurassic. The succession on the floor is as follows:-

            Pliocene
            Miocene
            Oligocene
            Eocene
            Jurassic

Very few exposures occur on the floors of the rifts.

Bores have proved the presence beneath the rift-valleys of a Permo-Triassic succession overlying a pre-Permian floor in the S. of the Upper Rhine graben; this consists of crystalline rocks similar to those of the Vosges.

The Jurassic has also been penetrated in boreholes, but is exposed only at the S. end of the Upper Rhine graben. The Lias consists of marls and shales, the M. Jurassic (Dogger) of oolitic and reef limestones and the U. Jurassic, preserved only in the S. of the Upper Rhine graben, of marls with reef and bioclastic limestones.

With the exception of the Lower Rhine graben the Cretaceous is missing, the Jurassic being overlain by thick and variable Tertiary successions. A summary will be given of the sequence in the Upper Rhine graben, well known for its important economic deposits. The pre-Tertiary floor, from N. to S., forms several basins separated by swells or horsts. The most significant of these are the Pechelbronn Basin (Eocene and L. Oligocene up to 800 m thick), the Strasbourg Basin (same formations, 1600 m thick), the Potassic Basin (same formations up to 2000 m thick) and the Mulhouse Horst (same formations only 200 m thick).

The type-succession occurs in the oil-producing Pechelbronn Basin, N. of Strasbourg. Eocene calcareous and dolomitic marls, with anhydrite and salt in places, are followed by Oligocene marls overlain by the oil-bearing sands, the cap-rocks of which are marls with salt and anhydrite/gypsum. The succession continues up into the Miocene consisting largely of clays and sands (further N. in the Mainz Basin carbonates occur).

Further S. the Potassic Basin contains a thick evaporite Oligocene succession with valuable sylvite deposits, as well as salt and anhydrite.

In the Mulhouse Horst the much thinner Eocene/Miocene succession is of a shelf type with limestones and sands and few evaporites. The Pliocene of the Upper Rhine graben, consisting of clays, sands and gravels, is generally

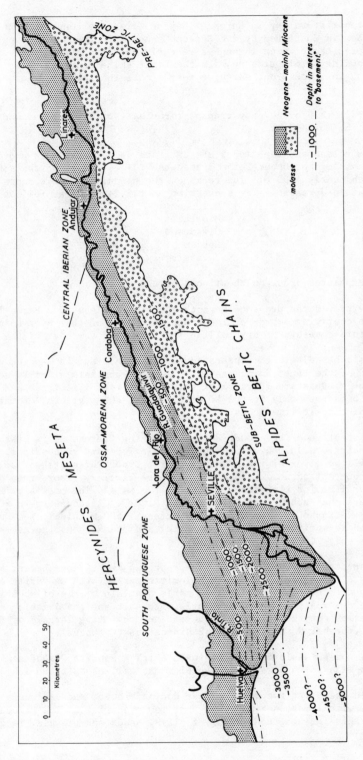

Fig. 8.4 STRUCTURE OF GUADALQUIVIR BASIN, SPAIN

less than 100 m but thickens to over 700 m N. of Heidelburg. Thick Quaternary deposits floor most of the valley. They consist not only of alluvium, but of loess up to 30 m thick.

Tertiary and Quaternary vulcanicity is discussed in Chapter 9.

Tectonic Evolution

Hercynian faults set the stage for the development of the Rhine rift-valleys; in particular N.N.E. faults (thought to be due to the interaction of the N.E. Variscan trend with a more fundamental northerly trend) were reactivated during foundering of the rifts.

At the end of the Cretaceous the Mesozoic rocks were fairly gently folded, initiating the basins which influenced Tertiary deposition. The first important faulting took place at the end of the Eocene. Those on the borders are on N.N.E. lines but in the interior N.N.W. faults are frequent. Nearly all are "normal faults", although on some there is evidence of transcurrent movement. These movements further developed the basins and thus influenced the important Oligocene sedimentation. At this stage the Hessen Graben was also formed, continuing the line of the Upper Rhine graben.

Further movements occurred in the U. Oligocene/L. Miocene. These included the development of a N.N.W. fault system extending to the Lower Rhine graben. Up till then the Upper Rhine graben had been in marine communication with the S., but the drainage now developed towards the N.; in any case the Hessen Graben had become blocked by the large Vogelsberg volcano. In the Pliocene differential uplift occurred (see also 5.6), and the rise of part of the Rhine Slate Mountains separated the Upper and Lower Rhine rifts.

Movements, particularly of the Upper graben, still continue, as shown by seismic activity and the survey of faults. There is a strong negative gravity anomaly.

8.3 The Ebro and Guadalquivir Basins (Spain) (Figs. 5.8 and 8.4)

Several Tertiary basins overlying the Iberian Hercynides are mentioned in Chapter 5, but the extensive and very deep Ebro and Guadalquivir require separate description.

The triangular Ebro Basin, 550 km in an E.-W. direction and up to 150 km wide, is framed by the Alpine fold-belts of the Pyrenees (7.15 - 7.19) and the Catalonian and Iberian Chains (7.20). Although low compared with the surrounding mountains much of the basin is around 400 m above sea-level. It is a dry region in which the Tertiary beds often form striking outcrops, particularly in the gorge of the Ebro at Fayon, downstream from Zaragoza.

Some excursions are described by Rios and Hancock (1961).

The basin is filled with a very thick Palaeogene succession overlain in places by relatively thin, molasse-type Miocene (Rios, 1961, 361). Triassic, Jurassic and Cretaceous have, however, been penetrated in deep bores; in places the Cretaceous is missing. In the N. of the basin the Eocene is marine; salt beds are present, those in the N.E. reaching depths of up to 3000 m below sea-level. In the S. continental Eocene was deposited; both facies are of considerable thickness. The rise of the Catalonian Chains at the end of the Eocene isolated the Ebro Basin from the open sea and the Oligocene deposits are continental, mainly lacustrine. The Marcilla bore penetrated 3415 m of Oligocene without reaching any older formation. In this N.W. part the basin is deepest; the base of the marine Eocene is at more than 4000 m below sea-level.

Tectonic Evolution

The position of the basin was probably determined during the development of Hercynian Spain with the establishment of a massif which, although subject to epeirogenic movements, was to resist further compression.

The basin as such, however, came into being with the late Eocene folding which formed the Alpine frame. As the Alpine Chains further developed in the Oligocene the subsiding basin was filled with its great thickness of Oligocene sediments. The palaeogene sediments are flat in the centre but become folded towards the margins; these folds pre-date the Miocene. Uplift which took place in the Pliocene was greater in the N. where the sediments are thickest, so that there is a tilt towards the S.

Along the N. margin movement of salt (halokinesis) has played an important tectonic role with the formation of diapirs.

In contrast to the Ebro Basin, the Guadalquivir Basin (Fig. 8.4) is a relatively narrow trench intervening between a great Alpine Chain - the Betic ranges (7.21 - 7.23) - to the S. and the Hercynian block or meseta to the N.; it is analogous, in fact, to the Saône-Rhône depression (8.1) in front of the Alps. The Guadalquivir Basin is mostly filled with Neogene sediments. The pre-Tertiary floor of the depression is up to 3000 m deep between Seville and the sea and geophysical results suggest that it descends offshore to depths of over 5000 m.

The Basin narrows and shallows E.N.E., where there are fairly numerous exposures of the Tertiary strata along the R. Guadalquivir.

Tectonic Evolution

The main development of the basin probably took place in the Miocene, with movements continuing into the Pliocene, as the Betic ranges rose to the S. The frontal or Prebetic zone of the ranges itself plunges under the Neogene of the basin towards the W.S.W.

The contrast between the two margins of the basin is matched by a sedimentary asymmetry; coarse molasse occurs along the high Betic front and finer along the edge of the meseta.

8.4 The Aquitaine and Paris Basins and the Low Countries (8.5 - 8.8)
    (France, Belgium, Holland, Denmark and West Germany)

Introduction

Between the Pyrenees and the Baltic, apart from the Armorican Massif (5.8), mainland Europe has a wide fringe of coastal plains and low plateaus underlain by Mesozoic and Tertiary. Strata of these ages, in fact, form large areas of Western France and Belgium, virtually the whole of Denmark and

Holland, and part of North-West Germany.

The Aquitaine Basin is bounded to the S. by Alpine thrusts along the N. face of the Pyrenees (7.16); to the E. and N. it has a Hercynian frame. To the W., not far from the coast, the continental slope begins, leading to the abyssal depths of the Bay of Biscay (Vigneaux, 1974, 275). Between the Pyrenees and the Montagne Noire it is connected through the Carcasonne Gap with the Rhône Corridor (8.1). The Jurassic floored Poitiers "door" connects the Aquitaine Basin with the Paris Basin; the structural division between the two basins is a horst, S.W. of Poitiers, bringing up a small inlier of granite.

The Paris Basin is also partly framed by outcrops of Pre-Permian rocks, affected by the Hercynian orogeny. There are, however, considerable gaps in the surface expression of this Hercynian frame, including the Poitiers "door" just mentioned. In the E., N. of the Vosges, the wide Permian outcrop of the eastern margin of the basin is in faulted contact with the Upper Rhine graben (8.2). To the S.W. the basin is connected by the Dijon "door" with the Saône-Rhône graben (8.1). To the N.W. the Cretaceous continues under the Channel to reappear in Southern England (8.8). To the N., where the Tertiary of the Paris Basin continues into Belgium, the Boulonnais-Artois Anticline is the limit of the Paris Basin.

North of this anticline, and of the Brabant Massif, Mesozoic and Tertiary strata extend to the S. margin of the Baltic Shield (2.2). This spread of younger strata spans five countries and is largely bounded by the sea. To the S.E. it merges into the Lower Rhine graben (8.2).

The following formations are exposed in the three regions (apart from the Paris Basin where there are small Palaeozoic outcrops):-

                        Pliocene
                        Miocene
                        Oligocene
                        Eocene
                        Palaeocene
                        Cretaceous
                        Jurassic
                        Triassic
                        Permian

Tectonic events (not equally developed throughout) affecting these formations were:-

Pre-Triassic	- formations affected mostly hidden
Late Triassic	- Old Cimmerian
Intra-Jurassic	- Saxonian and Young Cimmerian
Pre-Albian Late Cretaceous/Palaeocene	- Early Alpine
Eocene Oligocene Miocene	- Late Alpine

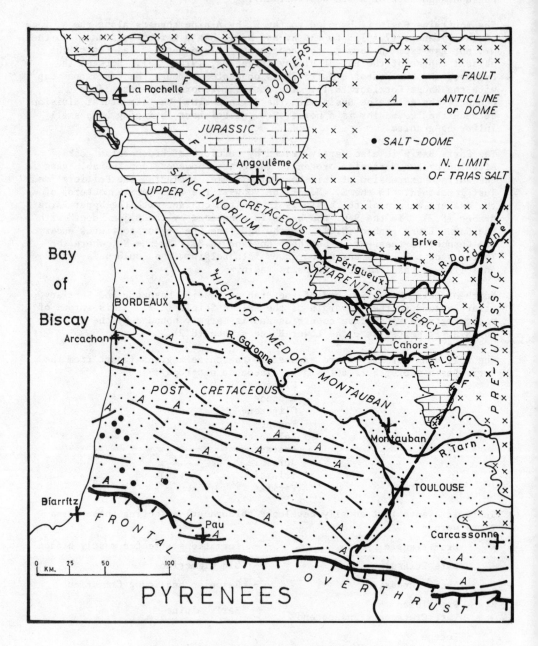

Fig. 8.5 STRUCTURE OF AQUITAINE BASIN
(partly after Vigneaux 1975)

8.5 The Aquitaine Basin (Fig. 8.5)

In the N.E. of the basin Jurassic limestones form plateaus, up to about 400 m, and rocky scarps overlooking the picturesque valleys of the Dordogne and Lot. The many cave-systems include the famous pre-historic sites of Lasceaux, Les Eyzies and Rocamadour.

In contrast, the centre and W. are low and flat with only scattered exposures; S. of Bordeaux lies the sandy plain of Les Landes, bordered by huge coastal dunes. The Parentis basin, under Les Landes, contains 80% of the known oil reserves of France; the Adour basin, near Pau, contains oil and abundant gas. The reservoir rocks are mainly Jurassic and Lower Cretaceous carbonates, more rarely sandstones.

Excursions in W. Aquitaine are described by Vigneaux (1975).

No pre-Permian rocks outcrop, but rocks older than the Permian have been reached in numerous bores. In the N. a Hercynian platform lies at relatively shallow depth, but in the S. the basin is more than 5000 m deep and more than 7000 m deep under Les Landes.

Under the N. of the basin Precambrian metamorphic rocks occur, probably Cadomian and probably the westerly continuation of the "ancient kernel" of the Massif Central (5.2). Between Bordeaux and Toulouse a number of bores have struck fossiliferous rocks ranging from Cambrian to Silurian. Devonian and Carboniferous have also been found; Stephanian Coal Measures have only rarely been reached (Landes).

The Permian outcrops only in the N.E.; around Brive red conglomerates and sandstones make conspicuous scarps. An extensive development has, however, been proved in bores between Périgeux and Toulouse.

The Trias is of widespread occurrence under younger strata and outcrops on the E. margin. In the deep southern part of the basin the full succession is developed, but in the much shallower northern hollow only Upper Trias occurs. The Lower Trias (Bunter) starts with a conglomerate followed by sandstones and silts. The Middle Trias contains shales with anhydrite and salt followed by carbonates (Muschelkalk). The Upper Trias (Keuper) in the S. consists mainly of shales with thick salt beds. These are thinner towards the N. of the southerly depression and in the shallow northern depression the Keuper becomes sandy.

In the Lias a shallow sea transgressed onto the Permian or older rocks E. and N. of Périgeux. The formation consists largely of shales and dolomites, but the occurrence of three major evaporite cycles (salt/anhydrite) shows that basins became cut-off. The Middle and Upper Jurassic contains thick, hard limestones, which form widespread outcrops in the E. and N.

The Lower Cretaceous outcrops only in the S. but has been proved to be thickly developed in the hydrocarbon-bearing basins (see above) of the Parentis and of the Adour. Estuarine-facies clays and clayey sands are followed by a dominantly calcareous succession, including porous oolitic limestone which is the oil-reservoir rock of the Parentis. The Upper Cretaceous forms a wide outcrop in the N. and N.E. where it overlaps, with only slight discordance, onto eroded Jurassic. It differs from the Chalk of the Paris Basin for it contains a considerable amount of continental material. The limestones are often clayey or sandy and shales and sandstones are present.

In the Lower and Middle Eocene shallow water marine or continental sedimentation took place on a northerly platform while a thin succession was deposited in the Parentis trough. Thick sedimentation took place in the subsiding sub-Pyrenean trough. The sediments include nummulitic limestones, marls and sandstones. On the platform between the Parentis and sub-Pyrenean trough carbonate sedimentation was dominant. From the Upper Eocene onwards the basin underwent very little tectonic disturbance. Molasse sediments, including fans of coarse gravels from the Pyrenees, were deposited in the S.E. and N.E., while marine limestones and marls were laid down in the W. The molasse/marine interface advanced and retreated as the Tertiary went on in response to minor epeirogenic movements. Red pebbly clays overlying the Miocene in places are regarded as Pliocene.

Tectonic Evolution

Alpine movements (sensu lato) played the main part in determining the structure of the Aquitaine Basin; pre-Triassic movements, however, determined the division of the basin into the shallow northern sector and the deep southern sector, separated by the Arcachon-Toulouse Anticline. In the northern sector the Saxonian or early Alpine folds are gentle and their N.W. trend shows the influence of underlying Armorican structures, as does the orientation of several faults cutting the Jurassic. In the S. the folds are more marked and the Pyrenean influence increases, especially in the sub-Pyrenean trough.

The discordance under the Upper Cretaceous is evidence of pre-Cenomanian compression; movements seem to have gone on throughout the early Tertiary and to have been marked in the early and late Eocene.

In the N. the Charentes-Quercy Synclinorium is followed to the S.W. by the Médoc-Montauban "high", for long a positive zone, where the Jurassic is thin and eroded; near Montauban, in fact, the Eocene rests directly on the pre-Permian. In the S. the folds, cut by thrusts near the Pyrenees, are complicated by "salt-tectonics" or halokinesis due to movement of the evaporites in the Upper Trias/Lias. Numerous diapirs have been formed, some breaking the surface, e.g. near Dax.

8.6 The Paris Basin (Fig. 8.6)

The Paris Basin, low-lying and flat in parts, is, nevertheless, diversified by platforms and undulating hills. Scarps, corresponding to the harder outcrops, are well developed towards the E. and S.E. The highest ground, for example that formed by Jurassic limestones N.W. of Dijon (around 800 m), is on the E. and S.E. borders, as are most of the good exposures; the Chalk, however, is well seen in numerous quarries and along high coastal cliffs, such as those near Dieppe. In the Tertiary areas natural exposures are few. The region escaped glaciation but permafrost effects are evident.

Excursions are described by Pomerol and Feugueur (1974).

Devonian and Carboniferous outcrops in the "Massif Palaeozoic de Ferques",

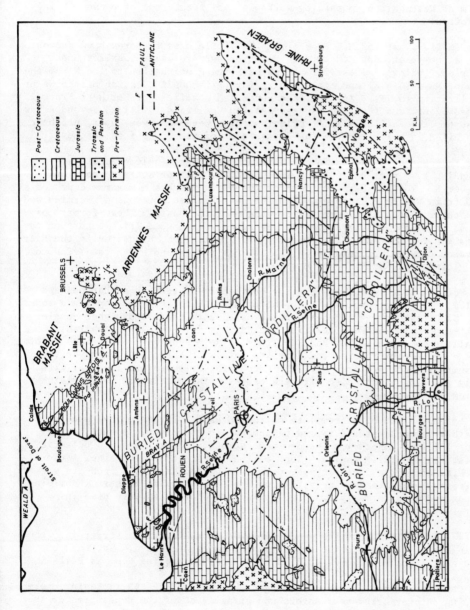

Fig. 8.6 STRUCTURE OF PARIS BASIN

only 28 km^2 in extent, on the Boulonnais Anticline. Givetian (M. Devonian) conglomerate and sandstone, overlain by limestone, are succeeded by Frasnian (U. Devonian) limestone and shale followed by Famennian sandstone (U. Devonian). Dinantian carbonates are succeeded by thin Namurian overlain by 200 m of Westphalian containing workable coal seams, not at present being exploited.

Near Caffiers, at the N.E. corner of the little massif, bores have shown that at no great depth the Middle Devonian unconformably overlies Upper Silurian (Ludlow). These formations mark the continuation to the W. of the Brabant Massif (4.6) and the Devonian/Carboniferous succession of the Namur Synclinorium (5.11). In fact, the concealed coalfield of Northern France extends westwards from the exposed Belgian Coalfield (5.11) to beyond Arras. Moreover, the Faille du Midi, bringing Lower Devonian over Upper, has been recognised in bores through the Jurassic S. of the "Massif de Ferques".

Near the centre of the basin oil bores have proved a belt of schists extending N.W. near Paris and a roughly E.-W. belt of schists and granite further S. The Pre-Permian floor of the basin reaches a maximum depth of 3000 ft in the Brie district, E. of Paris. Limited outcrops of Permian and wide outcrops of Trias form the E. margin of the basin. These formations extend in depth westwards to just beyond Paris and some Trias occurs on the W. margin, W. of Caen. Coal Measures of limnic type, deposited in the Saar-Lorraine depression, outcrop near Saarbruck and extend under the Trias to the vicinity of Nancy.

The Permian and Lower Trias consist dominantly of reddish sandstones and red shales; the Trias overlaps onto the granite, etc., of the Vosges and forms the "Sandstone Vosges" (Vosges grèseuses). Both basal and higher conglomerates occur. The Permian/Trias boundary is difficult to establish recalling the British New Red Sandstone. The Muschelkalk (M. Trias) is of marine or Germanic facies and consists mainly of carbonates; anhydrite/gypsum/salt beds are present. The Keuper (U. Triassic) is lagoonal with marly and sandy beds, also with evaporites. In Lorraine it is overlain by marine Rhaetic.

The extensive and often complete marine Jurassic succession has made it possible to work out a detailed palaeographic history, well outlined by Pomerol (1974, 235-239). The dominant strata are oolitic and bioclastic limestones and marly shales. The oolitic ironstones of Lorraine, littoral deposits of Middle Jurassic age, provide nine-tenths of the iron-ore worked in France. The fine-grained limestones of Caen furnish beautiful building stones and were formerly exported to England where they were used in Westminster Abbey and Canterbury Cathedral.

In the Lower Cretaceous the sea penetrated by the Burgundy strait into the Paris Basin, leading to the deposition of marls and clays. The sandy, continental Wealden facies (8.8) spread, however, as far S.E. as Paris.

The Cenomanian transgression submerged the whole of the Paris Basin, overlapping onto Armorica near Angers and bringing about the deposition of the Chalk.

Classic studies of the Palaeogene sands, limestones, marls (Palaeocene - Oligocene) and clays of the Paris Basin have proved five cycles of sedimentation resulting from a complex history of marine advance and retreat (Pomerol, 1974, 243-250).

Lower Miocene sands and gravels, brought by rivers from the Massif Central, are spread round Orleans. A mid-Miocene Atlantic transgression up the lower Seine Valley, as far as Blois, resulted in the deposition of sands and in the capture of the middle pre-Loire which had previously flowed towards the Seine. Pliocene deposits of limited extent are present.

Tectonic Evolution

Caledonian and Hercynian structures similar to those of the Brabant Massif (4.6) and the Namur Synclinorium (5.11) affect the Palaeozoic rocks of the Boulonnais Anticline (see above). Further S. the Bray Anticline brings up Upper Jurassic and is bordered on the N.E. by the Bray Fault which extends S.E. almost right across the Basin.

Further S. the Seine Fault, which passes Rouen, is one of a series of faults curving from S.E. to S. into the bisectrix of the two Hercynian Arcs and possibly forming the continuation of the Sillon Houiller system (5.2). From a study of the level of the Pre-Trias floor it is clear these faults date back to the Hercynian; from geophysical evidence they may even be Precambrian (Pomerol, 1974, 265).

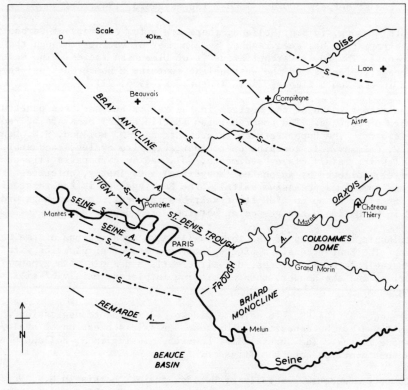

Fig. 8.7 STRUCTURE OF PARIS DISTRICT
(after Pomerol and Feugueur 1974)

Apart from these and other faults, the overall basin form is disturbed by numerous anticlines and synclines, the result of compressive phases of Upper Eocene/Lower Oligocene and Upper Oligocene/Miocene ages. Several occur around Paris itself, where they vary in direction (Fig. 8.7). Further N. and W., however, they have a dominantly N.W. strike and further E. a N.E. strike, evidence of the influence of the two Hercynian Arcs.

In the Pliocene, uplift took place resulting in the formation of platforms at several levels.

8.7 The Low Countries (Fig. 8.8)

This region mostly lies below 100 m and, apart from Belgium, is extensively mantled in glacial and more recent deposits; in fact, the topography is more often determined by glacial features than by pre-glacial formations, especially in Denmark. Solid rock outcrops are very few, and the low coastlines are eroded in drift deposits.

In Belgium the northerly sloping surface of the Lower Palaeozoic occurs at shallow depth under Tertiary, but in Holland it has not been reached in boreholes. In Denmark, on the other hand, a buried horst of Baltic Shield-type rocks has been struck in a bore at 1300 m below O.D. at the N. tip of Jutland; further S. another, similar horst, trending N.W., has been penetrated at between 800 and 1800 m (Noe-Nygaard, 1963, 2).

In South Limburg, in S.E. Holland, there are a few outcrops of Carboniferous and the formation has been reached in a number of borings through the Cretaceous. The Lower Carboniferous is of Dinantian facies. The Namurian is of clastic facies and the Westphalian contains a number of coal-seams, worked in the South Limburg field (Pannekoek, 1956, 21).

Permian is not known at the surface in the region but has been penetrated in a number of borings. The Lower Permian (Rotliegendes) consists of red clastic beds. The Upper Permian is present under N. Holland, S.W. Denmark and N.W. Germany. It contains up to four evaporite cycles, each starting with clastic, mainly clayey sediments, followed by carbonates (limestone and dolomite), followed by sulphates (anhydrite) and, lastly, chlorites (rock salt and sometimes potassium salts). The Rotliegendes is the reservoir rock, overlain by up to 4500 ft of salt-bearing beds, for the huge Groningen field in N. Holland, the gas being obtained from depths of 2600 to 3000 m.

The Triassic is made up of the three well-known subdivisions of the Germanic facies. The Bunter consists of conglomerates and other clastics with salt, the Muschelkalk of limestones, dolomitic marls and marls with gypsum/anhydrite, and the Keuper, thin or missing in places, mainly of siltstones with gypsum.

The presence of the Rhaetic marks a widespread marine transgression which was followed by the deposition of a long Jurassic succession of shale/limestone facies. The Jurassic is, however, missing in N. Holland where the Cretaceous overlaps onto the Triassic.

The Lower Cretaceous, generally of Wealden facies, occurs in N.W. Holland, W. of the ridge of Triassic which here comes near to the surface. The Lower Cretaceous is also present E. of this ridge in N.E. Holland, thickening into the Saxony basin of N.W. Germany. On the border of the two countries the

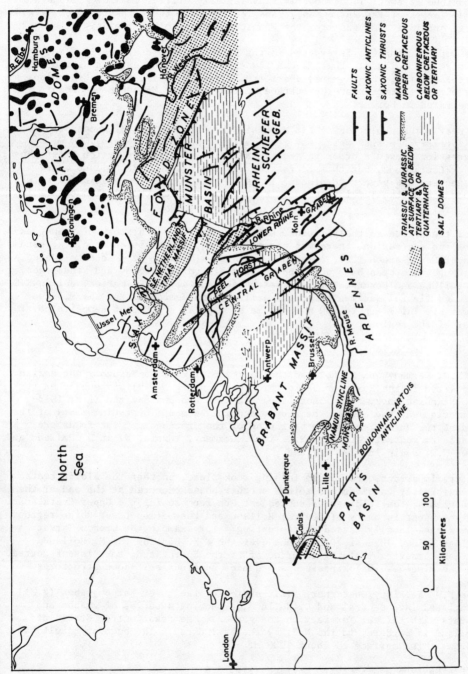

Fig. 8.8 STRUCTURE OF THE LOW COUNTRIES
(after Pannekoek 1956)

Schoonbeck field yields oil from gently folded Valangian (L. Cretaceous) sandstone at about 1000 m overlain by Tertiary drift. In N.W. Holland a sandy facies of the Barremian, transgressive onto the Triassic block, yields large quantities of gas from the Wanneperveen and other fields.

The Upper Cretaceous is exposed in the Munster basin and underlies much of Denmark and Holland; there are a few exposures in the extreme S.E. of the latter country. In Denmark, apart from Bornholm (2.1), the U. Cretaceous is the oldest formation at the surface, although older Mesozoic has been penetrated in bores.

In nearly the whole region the Mesozoic is overlain, mostly unconformably, by Tertiary deposits which may reach a thickness of more than 1000 m. Regression caused a break in sedimentation in much of the region after the Upper Cretaceous, but this regression was limited in S.E. Holland where Palaeocene has been recognised. It is in this region too that a few natural exposures of the Tertiary occur. Palaeocene (Danian Chalk) also occurs in Denmark.

From the Eocene to the Pliocene, although there was a history of transgression and regression, there was almost continuous deposition of clays, sands (sometimes glauconitic), silts and calcareous beds, mostly unconsolidated, although some beds have undergone diagenesis to limestones and sandstones. In Holland and Denmark Tertiary sedimentation was mainly marine, but in S.E. Holland fluvial sedimentation gradually became dominant from the Miocene onwards. Thick (up to 200 m in Denmark) glacial and Recent deposits cover most of the region.

Tectonic Evolution

In N.W. Germany and N. Holland the thick Upper Permian-Mesozoic succession deposited in the Saxonian basin (Richter-Burnburg, 1949) was subjected to the Saxonian movements. These took place in two phases, the first (Old-Cimmerian) at the end of the Keuper and the second (Young-Cimmerian) at the end of the Jurassic and the beginning of the Cretaceous. The folds are gentle to moderate striking N.W. in N. Germany, turning W. in N. Holland and then N.W. again in N.W. Holland.

Later, important, pre-Albian folding took place, so that the Albian rests unconformably on older Mesozoic. Another phase occurred at the end of the Cretaceous, the folding and subsequent erosion causing the Upper Cretaceous to be absent in parts of Central Holland and the German Lower Rhine region. This phase locally caused overthrusting; for example the Grondu Thrust brings folded Jurassic/Cretaceous over the N.E. edge of the N. Holland Triassic massif. Folding occurred after the Eocene and, to a lesser degree, in the Miocene. Halokinesis has produced numerous salt-dome structures.

The N.W. Tertiary/Quaternary fault-system of the Lower Rhine graben (8.2) continues into central and N. Holland, producing a series of graben and horsts with thicker Tertiary in the graben. The deep Central Graben of Holland is bordered to the N.E. by the Peel Horst. The boundary fault has a throw at the surface of about 1000 m.

8.8   Eastern and Southern England (Fig. 8.9)

The W. boundary of the region follows the base of the Permian from the E.

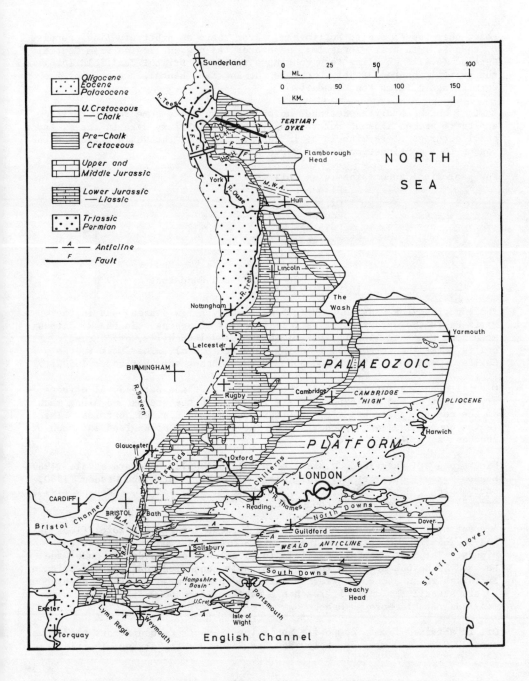

Fig. 8.9 STRUCTURE OF EASTERN AND SOUTHERN ENGLAND

coast near Sunderland to Nottingham. From there an arbitrary line, running S. then S.W. to near Gloucester, separates the present region from Central England (6.4). The line then continues round the Bristol Coalfield (6.6) to the Bristol Channel, S. of which, to the English Channel, the base of the Permian again forms the boundary.

East of the boundary as defined, no pre-Permian rocks come to the surface and there are no igneous outcrops apart from a Tertiary dyke in the N. and small outcrops of Permian lavas in the S. South of the Weald Anticline the region is a continuation of the Paris Basin (8.6).

The geological succession is as follows:-
>Pliocene
>Oligocene
>Eocene
>Palaeocene
>Cretaceous
>Jurassic
>
>Triassic)
>Permian   ) New Red Sandstone

The topography exceeds 1000 ft (305 m) in only a few places; as in the Paris Basin the harder outcrops form a series of scarps. In the N. Jurassic sandstone rises to 1489 ft (452 m) in the Cleveland Hills, and Jurassic oolite forms the Cotswolds, reaching 1070 ft (325 m). The Chalk in the Chilterns and in the North and South Downs reaches somewhat under 1000 ft.

Glacial deposits mantle much of the northern part and periglacial effects are evident in the S. Exposures are frequent in the scarps and along much of the coast; there are also numerous working and disused quarries. The region contains a large proportion of the population of England and much of the best agricultural land.

Excursions are described by Bisat et al. (1962, Hull), Blezard et al. (1967, London North), Curry et al. (1972, Isle of Wight), Curry and Wisden (1960, Southampton), House (1969, Dorset), Kirkaldy (1977, Weald), McKerrow (1973, Oxford) and Pitcher (1967, London South).

A great deal is now known about the pre-Permian (or Pre-Mesozoic) floor of the region from coal, gas, oil and water-bores and from geophysical techniques. The East Pennine Coalfield (6.4) extends eastwards far under the New Red Sandstone cover as the most valuable concealed coalfield in Britain. In Yorkshire and Lincolnshire, in fact, the pre-Permian floor has been shown to dip steadily E. until it reaches a depth of some 6000 ft (1830 m) below sea-level at Flamborough Head.

Oil is obtained from a composite anticline in Upper Carboniferous sandstone, some 2000 ft (610 m) deep, below New Red Sandstone at Eakring, 15 miles (24 km) N.N.E. of Nottingham.

South of the Wash the flat-topped London Platform is directly overlain by Mesozoic strata ranging up to the U. Cretaceous. The shallowest part appears to be an oval rise (-360 ft O.D.; 110 m O.D.) near Cambridge with a fall to about -1000 ft (-305 m) towards Harwich.

Folded Lower Palaeozoic rocks, frequently Silurian, make up much of the platform; around London Devonian strata occur, partly of Old Red Sandstone facies and partly of Belgian facies. At Oxford potentially valuable Coal Measures occur. In the S.E. (Kent Coalfield) the Coal Measures underlie strata ranging from the Lias to the L. Cretaceous at depths around 1000 to 1500 ft (305 - 458 m) below sea-level. The Measures generally rest on Carboniferous Limestone but to the N. overlap onto the Silurian.

The oldest of the exposed formations, the Permian, is characterised, as far S. as Nottingham, by the Magnesian Limestone, of marine origin. Between the Bristol and English Channels red, desert-type Permian sandstones occur, with basalt-lavas near Exeter. The Trias consists of the Bunter sandstones and conglomerates overlain by the Keuper sandstones and siltstones. However, the separation of the Permian and Triassic is by no means satisfactory and the whole sequence can be referred to as New Red Sandstone. Salt is obtained by pumping from the Upper Permian of the Middlesborough district.

The thin but widespread Rhaetic beds mark a major marine transgression. The Jurassic succession contains many of the classic stages. In fact, William Smith lived for a number of years in the region, near Bath, and made the observations which were fundamental in the development of stratigraphy.

Limestones and marls or shales normally make up the Lias, but a sandy, shallow-water facies was deposited over several tectonic "highs" including the Market Weighton (N. of the Humber), Bath and Mendip "Axes". The most conspicuous formations in the M. Jurassic are the oolitic limestones, much prized as building stones. In the U. Jurassic the Oxfordian and Kimmeridgian are dominantly argillaceous formations; the limestone of the Portlandian is another extensively quarried formation. Economically important bedded iron-stones occur in the L. and M. Jurassic.

The Jurassic succession is incomplete in several parts of its long outcrop, and the Cretaceous oversteps onto various subdivisions, as low as the Lias, above the Market Weighton "axis" and onto the Great Oolite near the S. coast. The movements responsible (see below) continued after the Jurassic resulting in both breaks and overstep in the L. Cretaceous sequence. The most complete succession is in the Weald Anticline, type-locality for the deltaic Wealden facies.

The Gault Clay and its facies variant the Upper Greensand (Albian) marks a widespread marine transgression, and are followed by the Chalk which reaches a maximum thickness of 1600 ft (485 m) in the S. East and S. of Exeter the Upper Greensand rests on the New Red Sandstone.

In the London and Hampshire Basins there are the most extensive outcrops in Great Britain of the Tertiary, resting unconformably on the Cretaceous. Broadly the widespread London Clay is overlain and underlain by dominantly sandy successions. In detail, however, a history of transgression and regression can be made out, as in the Paris Basin (8.6). In the Hampshire Basin, which includes the Isle of Wight, the Eocene is followed by an Oligocene succession.

In Norfolk and Suffolk "crags" (shelly, incoherent limestones, sands and clays) resting on the Chalk were formerly all assigned to the Pliocene but all except the oldest beds are now placed in the Quaternary.

Tectonic Evolution

The tectonic events which shaped the region (all Alpine, but influenced by older structures) were:-

Pre-Permian	-	formations affected now buried
Intra- and post-Jurassic	-	Saxonian
Lower Cretaceous) Late Cretaceous)		Early Alpine
Miocene	-	Late Alpine

The variations in facies in thickness across the "axes" referred to above indicates movements from the Lias onwards and, in the case of the Bath and Mendip "axes" at any rate, it is clear that much more ancient structures exercised control. There has, however, been controversy (e.g. Kent, 1956) whether these are true axes indicating anticlinal folding or whether they are stable blocks which resisted downwarping of their flanks.

Both folding and faulting occurred at the end of the Jurassic. In the Howardian Hills an intricate system of W. by N. and N.E. faults is of post-Jurassic and pre-Cretaceous age. Further S., a series of N.W.-S.E. aligned faulted synclines in the Upper Jurassic rocks runs out from beneath the Gault along the foot of the Marlborough and Berkshire downs and of the S. Chilterns (Arkell, 1947). In fact, the frequent unconformity at the base of the Cretaceous is a feature of much of the region. In many areas in Eastern England the Lower Greensand rests on the Kimmeridgian.

Crustal unrest with locally intense movements continued up to the pre-Albian interval. Post-Wealden/pre-Albian structures in the Weymouth area include the Poxwell Pericline, the Upton Syncline, the Ringstead Anticline and the Abbotsbury Fault-system; along the last-mentioned there was a southerly downthrow of as much as 1500 ft (457 m). On the upthrow side higher Jurassic and Wealden rocks were eroded prior to the transgression by the Albian sea. However, post-Cretaceous movement also took place along the fault-system. Regional uplift, slight warping and removal of varying amounts of Chalk preceeded the Tertiary.

Much of the structure of the region was due to Miocene movements. In the N. an easterly tilt was imposed off the Pennine Arch, modified in the Cleveland Hills where the Lias and Middle Jurassic were domed along an axis trending E. by S. Further S. the regional tilt becomes S.E. into the London basin which is complicated by a number of minor folds and several fractures. To the S. rises the classic Weald Anticline, or Anticlinorium, for it too is complicated by folds trending chiefly E.-W. and usually with steeper northern limbs to the anticlines (Fig. 8.10). Part of the Weald Anticlinorium may, however, have risen earlier (Lake, 1975, 551). The Weald is also cut by a number of faults trending generally E.-W. but turning E.S.E. towards the E. The minor folds and the faults within the London and Weald structures may be due to incompetent adjustment of the cover to deformation of the pre-Mesozoic floor.

The most intense effects of the Alpine Orogeny in Britain were felt along the extreme southern fringe of the area, that is from the Isle of Wight to the Weymouth Peninsula. Two major monoclines, the Sandown and Brixton

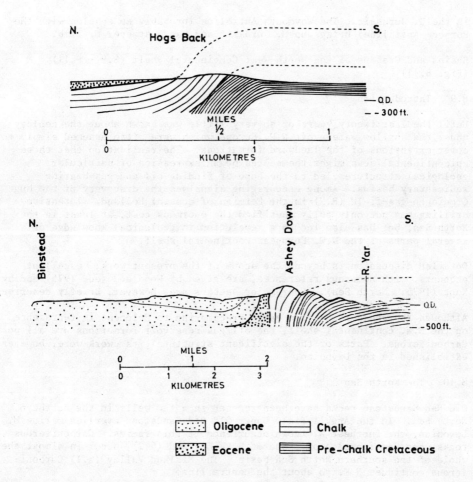

Fig. 8.10 SECTIONS OF SOUTHERN ENGLAND
A. Hogs Back, near Guildford.
B. Isle of Wight.

folds, affect the Isle of Wight (Fig. 8.10). They are arranged en echelon, and where each dies out (one westwards, the other eastwards) the outcrop of the Chalk widens appreciably (between Calbourne and Gatcombe). The northern limbs of both folds are almost vertical as revealed by the striking Chalk pinnacles of The Needles and the highly coloured Eocene Beds in Alum Bay, in the Isle of Wight.

This fold system extends W. at least as far as Weymouth, although it is probable that the individual folds are not continuous due both to en echelon development and cross-faulting (Donovan and Stride, 1961). On the steep N. limb of the Purbeck Anticline the Chalk is locally vertical. On the N. flank of this anticline, at Lulworth, E. of Weymouth, spectacular sharp folds occur in the U. Jurassic. The Weymouth Anticline (probably en echelon with the Purbeck Anticline) brings up M. Jurassic and has a steeper N. limb.

Basins and Grabens of the North-West Continental Shelf (8.9 - 8.11) (Fig. 8.11)

8.9 Introduction

Until the last twenty years or so very little was known about the geology under the shallow waters off N.W. Europe which were often assumed simply to cover extensions of the landward formations. The realisation that these epicontinental seas might themselves be an expression of particular geological structures led to the hope of finding oil and gas-bearing sedimentary basins. Among encouraging signs was the discovery of the huge Groningen gas-field (8.7) in the Permian of coastal Holland. Intensive drilling has not only fully justified the enormous cost, at least in the North Sea, but has also led to a revolution in geological knowledge of several parts of the N.W. European continental shelf.

Detailed discussion is beyond the scope of the present work; excellent accounts, with numerous references, are given by Woodland (ed. 1975) and by Kent (1975: North Sea). Four major sectors are, however, briefly described.

Although the whole range from early Precambrian to Holocene is represented on the N.W. continental shelf, the oil/gas reservoir formations are all post-Carboniferous. Parts of the significant structural framework were, however, established in the Devonian.

8.10 The North Sea

Old Red Sandstone rocks have been reached in a few wells in the northern North Sea. In the Argyll field Upper Old Red Sandstone overlies marine M. Devonian, the furthest N. known occurrence of this facies. Carboniferous rocks, the continuation of those of N.E. England (6.2), occur in almost the whole of the southern North Sea basin. The Midland Valley (6.1) Carboniferous continues N.E. to about the centre line.

The L. Permian continental sandstones of Rotliegendes facies (8.7) provide the main gas reservoir of the southern North Sea, the cap-rock being the Zeichstein evaporite sequence. The gas has migrated upwards from the Coal Measures. The Permian includes basaltic volcanics.

The Trias, consisting mainly of red, fine clastics (fine sandstones, siltstones and mudstones), underlies a large part of the North Sea. Off N. Denmark and S. Norway it is up to 3600 m thick and in the Viking Graben (see below) up to 1800 m. An almost complete Jurassic sequence occurs but continuing instability has caused many local breaks. In the northern sector estuarine/deltaic sandstones, belonging to the Rhaetic-Lias, M. Jurassic and U. Jurassic (Oxfordian), have been found to be excellent reservoir rocks, although it was not thought when exploitation began that the Jurassic would be important. Basaltic volcanics of M. Jurassic age occur in and near the Forties field between Peterhead and S. Norway.

The L. Cretaceous, mainly of argillaceous facies, tends to be a minor element in the stratigraphical column. The U. Cretaceous over most of the North Sea area is developed as Chalk. In the central area the Palaeocene

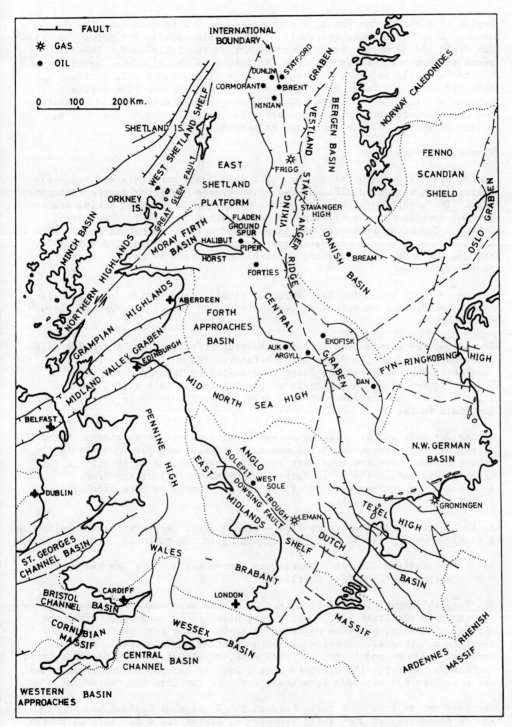

Fig. 8.11 BROAD STRUCTURE OF THE NORTH-WEST CONTINENTAL SHELF (partly after Kent 1975)

begins with the Danian Chalk, distinguished by its fossils from the Cretaceous. This is the main reservoir of the Ekofisk fields off Norway and the Dan field off Denmark. Sands occur higher in the Palaeocene. Deposition seems to have been continuous from the Eocene to the Pliocene, mainly of a monotonous shale sequence with sands in the Pliocene. A tuff horizon, referred to as the Palaeocene Ash Marker, is widespread. The Tertiary reaches a thickness of 3500 m near Ekofisk. In the Argyll field, beneath 500 m of Pleistocene, thicknesses are: Pliocene 610 m; Miocene 450 m; Oligocene 400 m; Eocene 430 m and Palaeocene 120 m.

Tectonic Evolution

In the broad structural sense the North Sea covers northerly and southerly basins (each with subdivisions) separated by the Mid North Sea/Fyn-Ringkøben High. These structures are cut across by the N.-S. striking Viking Graben between Shetland and Norway and its continuation to the S., the Central Graben. The broad framework was established with the breaking up of the Caledonian orogen into blocks including the Southern Uplands, continued by the Mid North Sea High, and the Wales-Brabant Massif at the S. end of the southerly basin.

In the southern basin the Carboniferous strata, deposited N. of the resistant Wales-Brabant Massif, were gently folded by the late Hercynian (Asturian) movements.

Throughout the Permian and Triassic the greater part of the North Sea area underwent subsidence, sedimentation being related to the separately developing N. and S. basins, largely separated by the Mid North Sea/Ringkøbing-Fyn High. Halokinesis of the Zeichstein salt started in the Triassic with the formation of salt diapirs, particularly in the North Sea gas field in the S.

In the Jurassic and L. Cretaceous taphrogenic control of subsidence became widely dominant and strongly influenced sedimentation, particularly in the Viking Graben. For the most part the U. Cretaceous sagged into the earlier depressions, but in the S. in later Cretaceous/early Tertiary times, inversion, i.e. upward warping of the earlier formed troughs, took place, probably due to compression.

In the Tertiary taphrogenic control ceased and the Tertiary basin developed as a simple downwarp of the whole area centring on the main rift system.

8.11 The English Channel, the Western Approaches and the Sea-Floor off North and North-West Scotland

The English Channel basin, which lies S. of the Hercynian front (passing almost under the Straits of Dover), originated as a trough soon after the Hercynian orogeny, as rocks of Permo-Triassic facies rest on eroded older rocks in mid-Channel. Subsidence in the central part continued in the Jurassic (1000 m) and in the L. Cretaceous, mainly of Wealden facies. Upper Cretaceous transgression extended over a partial barrier of Hercynian rocks which earlier had extended from Start Point to the Cotentin Peninsula.

Tertiary occurs in a wide basin between the W. tips of England and France and is dominantly of deep shelf limestones, except for a possible Oligocene emergence.

The structural history is analogous to that of S. England (8.8). Folding and faulting on E.-W. lines preceeded the Albian. Gentle warping took place in the Cretaceous/Tertiary interval and then there was post-Oligocene, probably Miocene, folding.

The Celtic Sea, off S.W. England and S. Ireland, merges rather imprecisely northwards into the Irish Sea. In the southern part of the latter a broad ridge of Precambrian and L. Palaeozoic rocks - the Irish Sea Geanticline - extends S.W. from the Lleyn Peninsula to the S.E. tip of Ireland. The ridge is, however, broken by two N. by E. graben, one not far off Lleyn and the other not far off S.E. Ireland. Between this and a shelf of L. Palaeozoic off the Welsh coast lies the Cardigan Bay Basin, floored by Mesozoic and Tertiary and continuing S.W. into the Celtic Basin, floored largely by Chalk and Palaeocene. At the N.E. end of the Cardigan Bay Basin a Permo-Triassic to U. Jurassic succession dips off the St. Tudwals Arch of "basement", off the S. end of Lleyn (Penn and Evans, 1976). East of the Arch, in Tremadoc Bay, this succession is overlain by Tertiary brought down against the L. Palaeozoic on land by the Mochras Fault, shown in the Mochras bore to have a downthrow to the W. of at least 4500 m. The inner Bristol Channel is floored by New Red Sandstone overlain further W. by Mesozoic about 2000 m thick in a syncline which is brought against the Palaeozoic of Devon by a major E.-W. fault just off the coast. The N. limb of the syncline is cut by a fault-zone which has a downthrow to the S. of about 1000 m. Further W. a Tertiary basin (5 boreholes proved M. to U. Oligocene) occurs E. of Lundy Island (Fletcher, 1975), cut off to the W. by the N.W. Sticklepath-Lustleigh fault, seen further S. in Devon. In fact, a number of the S.W. England faults continue into the outer Bristol Channel. Upper Cretaceous and Tertiary also occur W. of Lundy Island (Doré, 1976, 453).

North-west of the Irish Sea Geanticline there is a N.W. U. Palaeozoic to Triassic Basin.

The North Irish Sea (N. of Anglesey) is largely occupied by a Permo-Triassic Basin interrupted in the N. by the horst of the Isle of Man (4.3).

Pre-Albian movements forming E.-W. structures played an important part in shaping the structures, as in the English Channel. There was clearly post-Cretaceous/pre-Palaeocene disturbance, and both fold and fault movements at least as late as the end of the Palaeogene. Much of the faulting was N.N.E. to N.E. due to reactivation of Caledonian structures; for example, the Bala Fault of Wales (4.2) extends far out to sea.

The Permo-Triassic is saliferous and salt-dome structures have been formed. Gas has been proved in the U. Cretaceous off the S. Irish coast.

West of Shetland, and of the broad, submarine West Shetland Shelf (which may consist in part of Lewisian rocks, 2.3), a major N.E. fault-system brings down the large Faeroes/Shetland Mesozoic/Tertiary Basin. This is only partially explored but there are indications of the presence of oil. Much further out in the Atlantic, 450 km W. of the N. tip of the Outer Hebrides, a plateau of Mesozoic and Tertiary (overlying rocks of Lewisian type) surrounds the Rockall Tertiary Igneous Complex.

Between the West Minch Fault (following the S.E. coast of the Outer Hebrides) to the N.W. and Mainland Scotland and the Isle of Skye to the S.E., a basin of thick Mesozoic strata overlies probable Torridonian.

REFERENCES

Arkell, W.J. 1947. Geology of Oxford. Oxford.

Bisat, W.S. et al. 1962. Geology around the University Towns: Hull. Exc. Guide No. 11. Geol. Assoc.

Blezard, R.G. et al. 1967. The London Region (North of the Thames). Exc. Guide No. 30A. Geol. Assoc.

Brooks, M. & James, D.G. 1975. The geological results of seismic refraction surveys in the Bristol Channel. J.G.S., 131, 163-182.

Curry, D. et al. 1972. Geology of some British Coastal Areas: The Isle of Wight. Exc. Guide No. 25. Geol. Assoc.

Curry, D. & Wisden, D.E. 1960. Geology of some British Coastal Areas: The Southampton District. Exc. Guide No. 14. Geol. Assoc.

Debelmas, J. & Demarcq, G. 1974. Couloir Rhôdanien et le Bas Languedoc (articulation entre les Pyrenees, la Provence et les Alpes). In: Géologie de la France. Paris.

Donovan, D.T. & Stride, A.H. 1961. An acoustic survey of the sea floor south of Dorset and its geological interpretation. Phil. Trans. Roy. Soc., 244, 299.

Doré, A.G. 1976. Preliminary geological interpretation of the Bristol Channel approaches. J.G.S., 132, 453-459.

Fletcher, B.N. 1975. A new Tertiary basin east of Lundy Island. J.G.S., 131, 223-225.

House, M.R. 1969. The Dorset Coast from Poole to the Chesil Beach. Exc. Guide No. 22. Geol. Assoc.

Kent, P.E. 1956. The Market Weighton structure. Proc. Yorks. Geol. Soc., 30, 197.

Kent, P.E. 1975. Review of North Sea Basin development. J.G.S., 131, 435-468.

Kirkaldy, J.F. 1977. Geology of the Weald. Exc. Guide No. 29. Geol. Assoc.

Lake, R.D. 1975. The Structure of the Weald - a review. P.G.A., 86, 549-557.

McKerrow, W.S. 1973. Geology around the University Towns: Oxford district. Exc. Guide No. 3. Geol. Assoc.

Middlemiss, F.A. & Moullade, M. 1970. Summer Field Meeting in the South of France between Lyon and Avignon. P.G.A., 81, 303-361.

Noe-Nygaard, A. 1963. The Precambrian of Denmark. In: The Precambrian, Vol. 1. (ed. Rankama, K.). New York.

Nordmann, V. (ed.). 1928. Summary of the Geology of Denmark. Danmarks geologiske Undersøgelse V Raekke, No. 4. Keitzel, Copenhagen.

Pannekoek, A.J. (ed.). 1956. Geological History of the Netherlands. Government Printing and Publishing Office. 'S-Gravenhage.

Penn, I.E. & Evans, C.D.L. 1976. The Middle Jurassic (mainly Bathonian) of Cardigan Bay and its palaeogeographical significance. I.G.S. Rep. 76/6.

Pitcher, W.S. 1967. The London Region (South of the Thames). Exc. Guide No. 30B. Geol. Assoc.

Pomerol, Ch. 1974. Le bassin de Paris. In: Debelmas, J., Géologie de la France, 232-258.

Pomerol, Ch. & Feugueur, L. 1974. Bassin de Paris. Guides Géologiques Régionaux. Paris.

Richter-Burnburg, G. 1949. Anlage und Regionale Stellung des saxonischen Bechens. Erdol u. Tektonik (Amt. f. Bodenforschung, Hannover).

Rios, J.M. 1961. A Geological Itinerary through the Spanish Pyrenees. P.G.A., 72, 359-371.

Rios, J.M. & Hancock, J.M. 1961. Summer Field Meeting in the Spanish Pyrenees. P.G.A., 72, 373-390.
Smythe, D.K. & others. 1972. Deep Sedimentary Basin below Northern Skye and the Little Minch. Nat. Phys. Sci., 236, 87-89.
Vigneaux, M. 1974. The Geology and Sedimentation History of the Bay of Biscay. In: Nairn, A.E.M. and Stehli, F.G., The Ocean Basins and Margins. Vol. 2. The North Atlantic. Plenum Press, New York.
Vigneaux, M. 1975. Aquitaine Occidentale. Guides Géol. Rég. Paris.
Woodland, A.W. (ed.). 1975. Petroleum and the Continental Shelf of North-West Europe. 1. Geology. Inst. of Petroleum.

Chapter 9
TERTIARY AND QUATERNARY VOLCANIC STRUCTURES

9.1 Introduction

Although there are no active volcanoes in Western Europe, eruptions ceased in Central France only 3000 years ago and in West Germany about 10,000 years ago. In N.W. Britain there was widespread early Tertiary activity.

This vulcanism was controlled by tectonic events, both global and local; conversely the subsurface magmatism produced special structures, best seen in N.W. Britain where erosion has cut deeply into Eocene igneous complexes.

The Tertiary igneous rocks of Britain (9.2) occur mainly in several western Scottish Islands, in the Ardnamurchan Peninsula and in N.E. Ireland. Activity was associated with Atlantic rift and ceased about the end of the Eocene. In France vulcanism was widespread in the Massif Central (9.3) and extended S. to the Mediterranean. The activity associated with graben and basin formation started in the Upper Oligocene and went on until after the Wurm glaciation.

In West Germany (9.4) activity also continued until after the glaciations and is connected with the fault-system which formed the Rhine graben.

Outside of these three regions Tertiary or Quaternary volcanics occur in Western Europe at the S. end of the Western Alpine Arc (7.13) in France, and at the north-eastern end of the Catalonian Chain (7.20) and the eastern end of the Betic Chain (7.21) in Spain. These, and the Quaternary cone of Agde in Southern France, are westerly and now extinct volcanoes of the structurally-controlled and still active Mediterranean Belt.

9.2 The British Isles (Figs. 9.1 and 9.2)

Spreads of Eocene basalts occur in the Hebridean islands of Skye and Mull and in N.E. Ireland; they overlie Mesozoic and, more locally, thin Eocene sediments. Large Tertiary igneous complexes occur in Skye, Rhum, Ardnamurchan, Mull, Arran and N.E. Ireland (Richey 1961; Charlesworth et al. 1960; Charlesworth 1963). Many contain ring-dykes and cone-sheets, the recognition of which was an important advance in the understanding of the mechanics of igneous intrusion.

The Tertiary centres occur within a N.-S. zone, 250 miles (400 km) long and 40 miles (65 km) wide. They show few relationships to pre-Tertiary structures penetrating indifferently the N.W. Caledonian front, the Northern Highlands, the Grampian Highlands, the Midland Valley of Scotland and the Irish continuation of the Southern Uplands. Moreover, the igneous rocks all pre-date the Neogene topography and have been deeply dissected by Tertiary erosion and by Pleistocene glaciation. They form some of the most spectacular scenery of the British Isles, rising to 3309 ft (1010 m) in the gabbro peaks of Skye. Coastal and mountain exposures are excellent; excursions are described by Brown et al. (1969, Skye), Skelhorn et al.

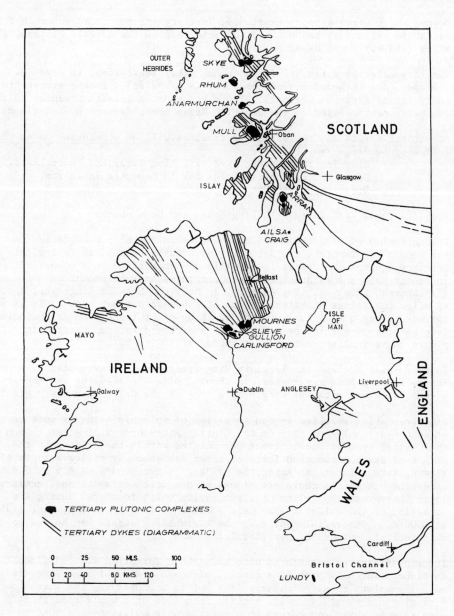

Fig. 9.1 TERTIARY IGNEOUS DISTRICTS OF BRITISH ISLES

(1969, Mull), Macgregor (1965, Arran) and Charlesworth & Preston (1960, Ireland).

Beyond the centres already mentioned, Tertiary igneous rocks occur in St. Kilda, 88 miles (141 km) W. by N. of Skye, and in the Rockall Plateau, 275 miles (440 km) W. of Skye.

Tertiary dolerite dykes are found, among other localities, in Mayo (W. Ireland), and in Snowdonia and Anglesey (N.W. Wales). Eocene granite forms Lundy Island (Bristol Channel). The dykes and the granite are near to the southerly continuation of the Scottish/Irish zone, (except N.W. Ireland).

Structural events due to Tertiary intrusions will be considered under the following headings :-
Ring-fractures associated with Ring-dykes and Cone-sheets.
Domes and Concentric Folds due to Magmatic Intrusion.
Dilation due to Dyke Intrusion.

Ring-fractures associated with Ring-dykes and Cone-sheets

The intrusion of ring-dykes and cone-sheets is not only a magmatic phenomenon but involves the formation of arcuate or annular faults. In the Tertiary Igneous centres it is possible to demonstrate in a number of instances both the existence of such fractures and the amount of displacement involved. For example, in Mull the Loch Ba Ring-dyke, about $4\frac{1}{4}$ miles in average diameter, is one of the most perfect known. It is intruded along a ring-fault, with downthrow of at least 3000 ft, bounding a late caldera of subsidence. The fault and ring-dyke are observed in a few places to be inclined outwards at angles of $70°$ to $80°$.

In north-east Ireland the Tertiary Ring-fracture of Slieve Gullion, up which has welled two acid ring-dykes, is about 7 miles in diameter and conforms in a remarkable way with the margin of a lobe of the Caledonian Newry granite.

Ring-fracturing resulted in the formation of volcanic calderas such as that of south-east Mull and that of the Central Ring-complex of Arran. In Mull the Central type of basalt lavas accumulated within the caldera to the extent of several thousand feet, at times (as shown by pillow-structure) flowing into a lake. In Arran the caldera is surrounded mainly by Old Red Sandstone but within there are blocks of Mesozoic sediments and Tertiary lavas interpreted as relics of a succession which foundered during the formation of the caldera. The main subsidence is estimated by King (1955) at 3000 ft. Sub-radial faults in the surrounding strata may be due to differential movement of peripheral segments.

The ring-fractures associated with cone-sheet formation are individually continuous for only part of a circle, although a zone of such fractures may make a complete ring. Similarly, the uplift accompanying the intrusion of a single cone-sheet is small, but the formation of a whole cone-sheet complex must have been marked by a considerable uplift of the contemporary surface. For example, on Creach Bheinn in Mull the cone-sheets total 3000 ft in thickness; as the average angle of inclination is about $45°$, the total uplift must have been over 4200 ft. Uplift of the same order also took place when the cone-sheets of Centre 2 in Ardnamurchan were intruded.

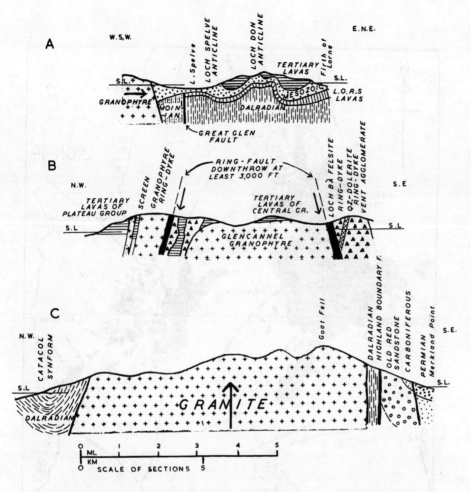

Fig. 9.2 SECTIONS SHOWING STRUCTURAL EFFECTS OF TERTIARY IGNEOUS INTRUSIONS.
A. Lateral compression - the concentric folds of south-east Mull.
B. Subsidence - the Loch Ba Ring - dyke.
C. Uplift - the granite dome of Northern Arran.

Domes and Concentric Folds due to Magmatic Intrusion

Major structural effects accompanied the emplacement of the Tertiary Granite Complex of North Arran. This has an almost perfect circular outcrop with a diameter of 8 miles, and everywhere, except on the N.E. margin, the surrounding schists dip steeply away from it, as do the Old Red Sandstone, Carboniferous, Permian and Triassic sediments, by which the schists are succeeded. Where the surrounding rocks originally dipped away from the granite, this dip has been merely accentuated and on the S.E. they have even been slightly overturned indicating outward pressure. On the N.W. side, the schists dip towards the granite and have been bent into a sharp syncline,

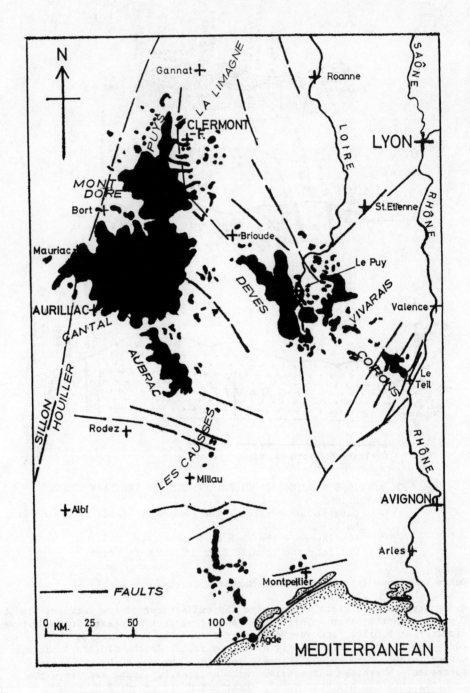

Fig. 9.3 TERTIARY AND QUATERNARY VOLCANIC DISTRICTS OF FRANCE

which is, in fact, a synform as it has been superimposed on Dalradian rocks
inverted by Caledonian folding. It is only on the N.E. side that the
granite seems to have broken through the schists, instead of carrying them
upwards with it, and here the schists strike sharply against the igneous
rock, rather than striking parallel with its margin. It is only on this
side too that the granite is in contact with rocks younger than the
Dalradian; at such contacts mylonite is developed. The granite dome, the
intrusion of which resulted in uplift of over 3000 ft, consists of two main
components, an older, outer, coarse-grained granite and a younger, inner,
fine-grained granite. Both granites are traversed by a zone of crushing
which may be the result of further upward pushing, with differential movement, after solidification.

Another remarkable tectonic effect of Tertiary intrusion is the concentric
folding peripheral to granophyres around the early caldera in Mull. The
folds describe about $270°$ of arc, within most of which there are two or
three well-marked anticlines with intervening synclines. The most striking
fold is the Loch Don Anticline, with a core of Dalradian schists, on the
eastern or outer limb of which the beds are locally vertical.

Dilation due to Dyke Intrusion

Most of the Tertiary complexes formed centres for the intrusion of great
linear dyke-swarms. The dykes are so numerous that there was significant
dilation of the continental crust at right angles to their general trend,
i.e. in a N.E.-S.W. or E.N.E.-W.S.W. direction (Richey, 1939). In the case
of the Arran swarm, for example, a belt of country which was originally
13.7 miles across in a north-easterly direction became 14.8 miles across
after dyke intrusion (21.9 km became 23.7 km).

Some of the dykes of the Mull swarm continue for remarkably long distances;
one dyke continues for 180 miles (283 km) to the N.E. coast of England,
cutting indifferently across older structures.

9.3   The Massif Central of France   (Figs. 9.3, 9.4 and 9.5)

The Tertiary and Quaternary igneous province of France extends for 310 km
southwards from N.W. of Clermont-Ferrand to Agde on the Mediterranean, and
for 250 km westwards from the W. margin of the Rhône graben to beyond
Mauriac. Guettard (1752) recognised the volcanic nature of some of the
mountains of Auvergne and the fact that these volcanoes break through
"fundamental rocks" helped to dispose of the Wernerian theory that they were
due to underground coal combustion.

Many of the cones form striking features superimposed on the topography of
the Massif Central 1000 m platform (5.1-5.2) (Rudel, 1970). Thus the Mont
Dore and Cantal Pliocene volcanoes rise respectively to 1885 m and 1838 m
above sea-level; the Cantal volcano has a diameter of 80 km (larger than
Etna). Others penetrate the Oligocene of internal rift-valleys and basins
or the Jurassic to the S. of the Massif. The most recent activity occurred
when the present topography of the Massif was virtually determined; ash
cones are perfectly preserved and flows occupy parts of valleys or cascade
over scarps.

Excursions are described by Roques et al. (1954) and Peterlongo (1971).

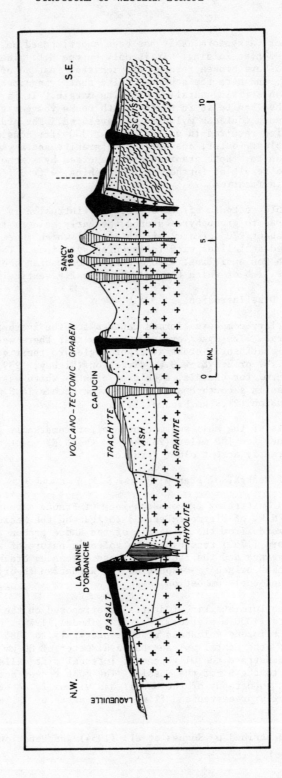

Fig. 9.4 SECTION OF MONT-DORE PLIOCENE VOLCANO (after Peterlongo 1972)

The structural associations of the pre-Pliocene volcanoes will first be described and then those of Pliocene and Quaternary volcanoes.

Structures Associated with Pre-Pliocene Volcanoes

The W. boundary fault of La Limagne, the largest Oligocene graben within the Massif (5.2), localised vulcanicity from the Upper Oligocene onwards. Among the earliest volcanoes are the two basanite flows interbedded with Burdigalian of the Plateau de Gergovie, which occur immediately W. of the fault. Others, of Upper Oligocene and of Miocene ages, occur both along the fault and within the graben.

The Miocene also saw the first eruptions of the Mont-Dore and Cantal volcanoes, situated on the most important volcanic zone of the Massif, which extend on a slightly sinuous N.-S. course for 310 km from N.W. of Clermont-Ferrand to the Mediterranean. Local tectonic control of much of the activity on this zone is clearly demonstrable (see below), but the zone itself must overlie an elongated, magma, dominantly basaltic, reservoir, for which a deep-seated lineament probably acted as a feeder.

Structures Associated with Pliocene and Quaternary Volcanoes

The Pliocene was the time of most widespread activity. The two great strato-volcanoes of Mont-Dore (Glangeaud et al., 1965) and Cantal were largely formed in the Pliocene; both are localised on sunken strips within the main N.-S. zone. The Mont Dore volcano, 80% pyroclastic, in addition to basalt material, includes a high proportion of acid material, both extrusive and intrusive; there was considerable movement of the centres of activity. Major eruptions took place in the Villefranchian (Lower Pleistocene). The Cantal volcano, simpler and made up of basalts, andesites and pyroclastics, may have become extinct at the end of the Pliocene (de Goër de Herve, 1973).

Most geologists would regard the Chain of Puys (Bentor, 1955), stretching N. and S. from the viewpoint of the trachytic Puy de Dôme, as the outstanding spectacle of the Massif Central. Some 80 cones, dominantly basaltic, are aligned on a N.-S. basement anticline, (perhaps raised by the underlying magma reservoir), which foundered and cracked.

It is structurally interesting that in detail small groups of puys lie on lines slightly oblique to the Chain as a whole, probably following fissures. Many of the basalt cones date from 13,000-10,000 years ago; great ash showers occurred about 8300 years ago and some volcanoes, like the Puy de Pariou, are slightly younger; the trachytic explosion of Lac Pavin took place only 3000 years ago, or even less.

Some of the flows spill over the Limagne fault. The fracture also localised Quaternary activity, notably that of the large Gravenoire ash-cone N. of Clermont-Ferrand.

On the main zone, S. of Cantal, the basalt volcanoes of Aubrac and the 200 basalt vents which stretch from the Causses to the Mediterranean are mainly Pliocene to early Middle Pleistocene.

Localisation by faults can be demonstrated, and in the most southerly Chain there was a displacement in time southwards, the youngest volcano being that

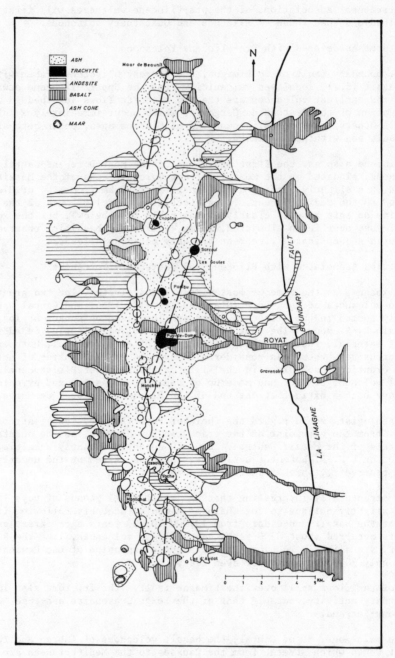

Fig. 9.5 CHAIN OF PUYS AND THEIR STRUCTURAL RELATIONSHIPS
(after Peterlongo 1972)

of Agde on the Mediterranean. The Sillon Houiller (5.2) also acted as a conduit, notably for the columnar phonolite of Bort.

Many of the late Miocene to late Palaeolithic volcanoes of the Vélay and Vivarais are associated with faults, and with the faulting and buckling of the Oligocene basin of Le Puy. In the ash of a Quaternary cone near Le Puy the skeleton of a man was discovered.

The most easterly flows, the Pliocene basalt of Coirons, overlook the Rhône graben and the Rochemaure necks are situated on one of the bounding post-Oligocene faults.

The geothermal gradient in the Massif Central is well above the average for Western Europe, and there are numerous warm and carbon dioxide springs.

9.4 The Rhineland (Fig. 9.6) (West Germany)

Tertiary and Quaternary volcanoes occur in the Rhineland in association with the rift-valley system; many are superimposed on the Schiefergebirge, S.E. of Bonn.

As in the Massif Central (9.3), activity started in the late Oligocene and continued well into the time of man; post-glacial ash cones, particularly in the Eifel, are similar to the French puys. Some excursions are described by Tilmann (1938).

Structures Associated with Pre-Pliocene Volcanoes

Uplift of the Rhineland (possibly due to regional magma emplacement) with subsidence of the rift-valleys began in the Cretaceous, and fault movement, as shown by the great thickness of sediments, was active in the Oligocene. The faults provided channels for the uprise in the Upper Oligocene of basaltic and more acid magma of alkaline type. The Siebengebirge volcanoes (on faults at the S.E. end of the Lower Rhine Basin), trachyte and phonolite plugs in the Westerwald (at the end of a small graben branching off the N. end of the main graben) and similar plugs in the Eifel, all date from this time. Activity increased in the Miocene, during which, in the Westerwald, basalt was intruded as dykes and spread over the surface as large sheets. The Swabian Plateau, E. of the Black Forest, is penetrated by a swarm of tuffisite pipes or diatremes (Cloos, 1941) and by the Steinheim and Ries explosion hollows of Upper Miocene age. These lie on a line parallel to the S.E. boundary of the Rhineland upwarp. Their origin is discussed by Holmes (1965, pp. 270 and 1052).

The loess-covered Kaiserstuhl (Wimmenauer, 1962), N. of Mulhouse, is the only volcano on the floor of the Upper Rhine graben (for discussion see Holmes, 1965, 1051).

Structures Associated with Pliocene and Quaternary Volcanoes

In the Pliocene and Quaternary the Rhineland, including the Schiefergebirge, slowly rose with faults separating differentially moving blocks. The large Vogelsberg volcano lies across the Hesse rift, the N.N.E. continuation of the main graben.

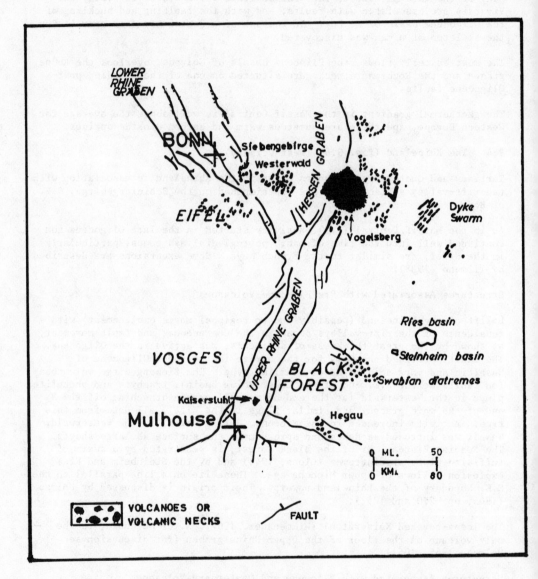

Fig. 9.6 TERTIARY AND QUATERNARY VOLCANIC DISTRICTS OF RHINELAND, WEST GERMANY

Quaternary activity in the Eifel is associated with fractures which are the S.E. continuation through the Schiefergebirge of the fault-system on the S.E. margin of the Lower Rhine Basin. Thus basaltic cones lie on the boundary of a small graben near Niedermendig which contains Tertiary sediments. Activity in this area included the emission of phonolitic tuffs and of trachytic tuffs, some of which was borne eastwards across the Rhine to the Westerwald. The tuffs date back 11,000 years. The Laacher See, N. of Niedermendig, is a lake-filled explosion vent punched through these ashes, about 2 km in diameter. This, and other crater lakes, (Erlenkeuser et al., 1970), are probably the youngest volcanoes of the Rhineland.

The geothermal gradient in much of the Rhineland is two or three times the average for Western Europe and there are numerous hot springs. Effervescing springs, rich in carbon dioxide, are also plentiful and are the sources of the mineral waters of the Rhineland spas.

REFERENCES

Bentor, Y. 1955. La chaine des Puys. Bull. Serv. Carte géol. France, 242.
Brown, G.M. et al. 1969. The Tertiary Igneous Geology of the Isle of Skye. Exc. Guide No. 13. Geol. Assoc.
Charlesworth, J.K. et al. 1960. The Geology of North-East Ireland. P.G.A., 71, 429-459.
Charlesworth, J.K. 1963. Historical Geology of Ireland.
Charlesworth, J.K. & Preston, J. 1960. Geology around the University Towns: North-East Ireland. The Belfast Area. Geol. Assoc.
Cloos, H. 1941. Bau und Tatigkeit von Tuffschloten. Geol. Rundschau, 5, 32, 709-800.
Erlenkeuser, H. et al. 1970. Neue C^{14} Datierungen sum Alter der Eifelmaare. Eiszaitalter Gegenw., 21, 177-181.
Glangeaud, Ph. et al. 1965. Géochronologie et évolution palaeomagnetique, géochimique, petrologique et tectonphysique du Massif volcanique de Mont-Dore. Bull. Soc. géol. Fr., (7), 8, 1000-1025.
Göer de Herve, Alain de. 1973. Quelques aspects du quaternarie dans les régions volcanisees du Massif Central. Ann. Sci. de l'Université de Besancon.
Guettard, J.-E. 1752. Mémoire sur quelques montagnes de la France qui ont été des volcans. Mém. Ac. des Sc.
Holmes, A. 1965. Principles of Physical Geology. rev. ed. Nelson, London.
King, B.C. 1955. The Ard Bheinn area of the Central Igneous Complex of Arran. Q.J.G.S., 110, 323.
Macgregor, M. et al. 1965. Excursion Guide to the Geology of Arran. Geol. Soc. Glasgow.
Peterlongo, J.M. 1972. Massif Central. Guides Géol. Rég. Paris.
Richey, J.E. 1939. The dykes of Scotland. Trans. Edinburgh Geol. Soc., 13, 393.
Richey, J.E. 1961. British Regional Geology: Scotland: The Tertiary Volcanic Districts. 3rd ed. M.G.S.
Roques, M. et al. 1954. Summer Field Meeting in the Massif Central, 1951. P.G.A., 65, 278-312.
Rudel, A. 1970. Les volcans d'Auvergne. Editions Volcans. Clermont-Ferrand.
Skelhorn, R.R. 1969. The Tertiary Igneous Geology of the Isle of Mull. Exc. Guide No. 30. Geol. Assoc.

Tilmann, N. et al. 1938. Summer Field Meeting, 1937: The Rhenish Schiefergebirge. P.G.A., 49, 225-260.
Wimmenauer, W. 1962. Zur Petrogenese der Eruptivgesteine und Karbonatite der Kaiserstuhls. Neues Jarbuch. Mineral Abhandl., 1962, 1-11.

Chapter 10
SUMMARY OF THE STRUCTURES OF WESTERN EUROPE AND THEIR SIGNIFICANCE IN PLATE TECTONICS MODELS

10.1 Distribution and Inter-relationship of Western European Structures (Fig. 10.1)

Precambrian complexes (1000 to 2800 m.y. old, some likely to have been derived from still older Katarchean rocks) probably make up much of the Western European continental crust. Nevertheless, they form only two continuous, large outcrops, namely the Baltic Shield (2.1) and its continuation into Southern Norway (2.2), and N.W. Scotland (2.3). Moreover, only in these two regions is it possible to study the older Precambrian unaffected or merely slightly affected by later events. These two regions are now separated by a wide branch of the Caledonian orogen, but it seems likely that in Archean times they formed part of a huge pre-Atlantic continental nucleus and shared a common history (cf. Pregothian (2.1) and Scourian (2.3)). In the Proterozoic, however, the history of the two regions became increasingly different as they probably moved further apart.

Much of the western edge of the Baltic Shield is made up of the northerly-striking Svecofennides, cut across in the Arctic by the roughly contemporaneous Karelides. In the S. the widespread Sveconorwegian metamorphism took place about 1000 m.y. ago.

In N.W. Scotland the N.E. orientated Scourian grain is cut across and partly replaced by the mainly N.W.-striking Inverian (2200 m.y.) and Laxfordian (1500 to 1600 m.y.) structures.

The Baltic Shield rocks and structures continue within the Caledonides at least as far W. as the Norwegian coastal islands (3.8) and have been proved under N. Denmark (8.7). Further S. they must also underlie parts of the Hercynides, as shown by the Icartian (2620 m.y.) and Penteverian (1000 to 1200 m.y.) gneisses of N.W. France and the Channel Islands (5.8). Here the Archean and Proterozoic have also been affected by the much later Cadomian orogeny (600 m.y.). This was probably extensive in the southern part of Western Europe as crystalline rocks, which may be of this age, have been proved under a large part of the Paris Basin (Fig. 8.6) and come up in the Hercynian Vosges and Black Forest (5.5), Massif Central (5.2) and Iberia (5.12). In Spain and Portugal, moreover, the occurrence under the Hercynides of much older Precambrian is evident.

The largely concealed kratogen under the English Midlands (6.4), and its better exposed western border, provide no definite evidence of the presence of rocks older than the Upper Proterozoic. In the Armorican Arc of S.W. England (5.9) some metamorphic rocks of possible Cadomian age are present. Within the Caledonides, to the N. of the Arc, the break between the late Precambrian and Cambrian in South Wales (4.2) is probably due to Cadomian movements.

The folding and metamorphism of the Mona Complex (4.2), in North Wales, may have been caused by an earlier Cadomian phase, but this must have been somewhat before the end of the Precambrian as the top of the Precambrian near

Bangor (4.2), not far to the E., passes up into the Cambrian. No Precambrian occurs in the Armorican Arc in Ireland (5.10), but the Rosslare Complex (4.5) proves the presence of Archean or Lower Proterozoic under the Caledonides.

The Caledonides of the Northern Highlands (3.3), and of Shetland (3.6), which certainly lie N. of the Caledonian suture (10.2), have a Lewisian floor. Further S. this "basement" does not outcrop (although it may be present in depth) in the Scottish Grampians (3.4) but comes up in the W. of Ireland (3.4). It may also be present under the Midland Valley and at least the northern part of the Southern Uplands.

Between the North-West Caledonian Front (3.2) and the Church Stretton line (4.2), and N. of the Hercynian Front, the British Isles form part of the Caledonian orogen. From the North Sea coast of Scotland to central Ireland the Highland Boundary Fracture-Zone separates rocks which have undergone Caledonian metamorphism to the N.W. from those in which metamorphism has not gone beyond slaty-cleavage. Near the Atlantic coast of Ireland, however, regionally metamorphosed rocks occur S. of the Fracture-Zone (3.5). South of the latter, too, large sections of the Caledonian fold-belt are hidden under younger strata.

The other major element of the Caledonides in Western Europe forms nearly the whole of Norway and Sweden (3.8), W. of the eastern Caledonian front (3.7). Metamorphic and non-metamorphic regimes are not as clearly separated as in the British Isles. Foreland folding is also much more strongly developed E. and S. (2.1 and 2.2) of the Scandinavian front than that W. of the N.W. front in Scotland.

The non-metamorphic Caledonian fold-belt probably continues from Wales (4.2), round the Midlands kratogen (6.4), under London (8.8) and on to the Brabant Massif (4.6). This implies that there is no continuation between the S.E. Caledonian Front (the Church Stretton line) in Britain and that in Norway, and that a considerable part of the Caledonian fold-belt deeply underlies Holland, most of Denmark and northern Belgium. The Norwegian front may curve eastwards to the Sudetian Caledonides. Eastwards continuation of the Caledonides through central Europe is also supported by the fairly certain presence of Caledonian metamorphic rocks in the Alps (7.1).

South of the Hercynian front Caledonian orogenic elements form the Ardennes, but are absent in Armorica and only doubtfully present in S.W. England. Further S., in the Massif Central (5.2) folding and metamorphism of Caledonian date are doubtful, and Caledonian metamorphism (and probably also folding) is certainly absent in the Montagne Noire (5.3). There is no evidence of a Caledonian metamorphic or structural event in the Pyrenees (7.15) or in the Iberian Hercynides (5.12), although a break at the base of the Ordovician in the Central Iberian Zone (5.14) shows that some Palaeozoic disturbance took place.

It would seem, therefore, that in the southern part of Western Europe there was no Caledonian metamorphism and only minor Caledonian folding. In fact, there is only limited overlap of the Cadomian orogenic belt and that of the Caledonian.

A Hercynian front, marking the N. limit of metamorphism and intense folding, can be traced from the Rhine to S.W. Ireland. Broadly, the most altered

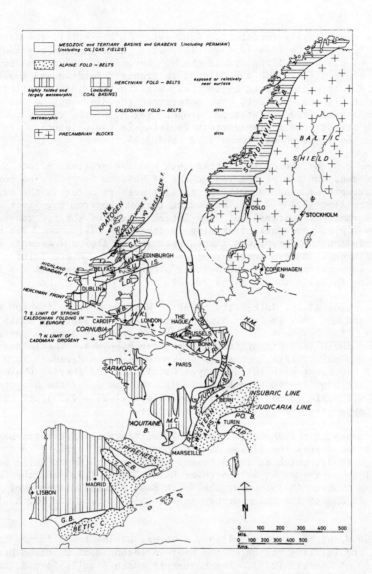

Fig. 10.1 GEOLOGICAL STRUCTURE OF WESTERN EUROPE
A. = Ardennes; A.P. = Apennines; B.F. = Black Forest; B.M. = Brabant Massif; C. = Connemara; C.C. = Catalonian Chains; C.G. = Central Graben; E.B. = Ebro Basin; G.B. = Guadalquivir Basin; G.H. = Grampian Highlands; H.M. = Harz Mountains; I.C. = Iberian Chains; I.S. = Iapetus Suture; L.D. = Lake District; L.R.G. = Lower Rhine Graben; M.C. = Massif Central; M.K. = Midlands Kratogen; M.V. = Midland Valley; N.H. = Northern Highlands; O.G. = Oslo Graben; R.C. = Rhône Corridor; R.S. Rheinisches Schiefergebirge; S.G. = Saône Graben; S.U. Southern Uplands; U.R.G. = Upper Rhine Graben; V. = Vosges; V.G. = Viking Graben.

rocks are considerably further S., in Iberia (5.12 - 5.14), the Massif Central (5.2) and the Alps (7.1 - 7.14), with the metamorphic grade and the presence of plutons falling off to the N. Beyond the Hercynian Front, however, (in contrast with the Caledonian fronts), folding and faulting persists for several hundred kilometres, e.g. to the Midland Valley of Scotland (6.2), and, in fact, have determined the distribution and structure of the main coalfields of Western Europe. Far N. of the front, too, strong Hercynian stress caused transcurrent faulting in the Caledonian blocks as far N. as the Scottish Highlands (3.3 and 3.4).

The Alpine orogeny, being the latest, has had the greatest effect on the structure and, therefore, the topography of Western Europe. The most complex structures and the most spectacular scenery occur in the huge arc of the Western Alps (7.1 - 7.14), extending for 600 km from the Riviera to the eastern borders of Switzerland. The internal part of the arc was formed by the early Alpine movements, finishing in the early Oligocene. The external part, including the Juras, was formed by the late Alpine movements going on to the Pliocene. In Provence (7.13), however, E.-W. Eocene folding dominates and interesting interference folds occur where these meet the N.-S. late Alpine folds in the Dauphiné (7.10).

Two other major Alpine fold-belts occur in Western Europe, the 400 km long Pyrenees and the 1000 km long, high Betic Chains in Southern Spain and the Balearics. The Pyrenees (7.15 - 7.19) differ from the Alps in the rarity of nappes and the higher proportion of Hercynian elements. The fold-structure is E.-W. and mainly late Eocene, as in Provence. The Pyrenees form a triangular frame along with the Jura-type Catalonian and Iberian Chains (7.20) for the Tertiary floored Ebro Basin (8.3). To the S. of the Neogene Guadalquivir Basin (8.3), the Betic Cordillera (7.21 - 7.23) are characterised by nappe-structures of early Miocene age.

From the mouth of the Rhone (and probably out into the Mediterranean) to the S. end of the Vosges the Alpine Arc is bordered by a Tertiary tectonic depression (8.1), which at its N.E. end joins up with Upper Rhine Graben (8.2). This continues N. as the Lower Rhine Graben (8.2) and the Central and Viking Grabens (8.10) under the North Sea, a total distance of 1950 km from the S. end of the Rhine Graben and of 2550 km from the mouth of the Rhône.

Far to the W. of these rift-valleys, however, Alpine folds formed the Mesozoic and Tertiary basins of the western seaboard of mainland Europe (8.5 - 8.7) and of England (8.8). The movements which finally formed these basins were Tertiary, but breaks in the sequences show that there were areas of instability (e.g. the Saxonian movements) throughout the Mesozoic.

Alpine folding and faulting also affected the structures of Central England (6.4) and Northern England (6.3) and, to a lesser extent, those of Scotland and Ireland.

10.2 Plate Tectonics Models (Figs. 10.1 and 10.2)

The concept of mountain formation by crustal compression arose largely from the study of European structures, and by the end of the 19th century it was accepted that the crumpling of the Alps resulted from the approach of Africa and Europe. The demonstration of such global movements led on to Holmes and others postulating the movements of continental blocks by convection

currents. From this theory and the study of seismic disturbances there
developed the concept of Plate Tectonics.

To explain European structures, increasingly sophisticated models, some
involving interacting "microplates", have been erected, dismantled and
replaced. To evaluate these would require another volume; what follows is
an attempt to fit the major structures described in previous Chapters into
some broad plate frameworks.

The major plates involved in the formation of European structures from the
Mesozoic up to the Present are the Eurasian Plate, the American Plate and
the African Plate. The same plates, with different boundaries, also inter-
acted to produce European structures at least as far back as the late
Precambrian.

At the beginning of the Mesozoic the African and Eurasian Plates lay far
apart separated by the broad, wedge-shaped (widening eastwards) Tethys Sea.
The African Plate probably lay further W. relative to Europe. Throughout
most of the Jurassic and Cretaceous the two plates approached each other and
the African Plate moved E. past the European.

This latter movement broke off and then rotated microplates, of which the
largest are those of Iberia, Apulia (part of Italy) and Turkey. Up to 18
smaller microplates have been postulated by Dewey et al. (1973), although
there is as yet little palaeomagnetic evidence for many of these (Ager,
1975, 147). On the other hand, there is strong geological and palaeomagnetic (van der Voo, 1969) evidence for the anticlockwise rotation of Spain
and probably also for its eastwards slip along the North Pyrenean Fault
(7.16), crushing and folding the Pyrenees as it did so.

The spasmodic closing of the great jaws of Africa and Europe resulted in
major tectonic episodes in the early Tertiary which formed the External
Chains of the Western Alps. The Insubric Line is probably the suture.
Oceanic crust was heaved up, notably in the Italian Alps N. of Turin. The
great arc of the Western Alps may be due to the Carnic microplate driving
the folds into the gap between the Bohemian and Central France Massifs.

A last major push in the Neogene, going on into the early Pliocene, formed
the External Chains and thrust parts of their N. and N.W. margin over their
own erosional debris (7.2).

The energy of the converging plates was transmitted at intervals from at
least the early Jurassic to the Miocene, far to the N. and N.W., folding and
faulting the basins of the W. seaboard of mainland Europe (8.4 - 8.7), of
the North Sea (8.10) and of S.E. England (8.9). In Britain marked folding,
probably Miocene, did not extend much further N. than the Bristol Channel
(5.9). Further N. the main tectonic influence was the opening of the
northern part of the Atlantic, itself of course part of the main plate
movements. This part of the Atlantic was still narrow some 60 m.y. ago
(Harland, 1969) (a view supported by palaeomagnetic evidence) when intense
igneous activity (9.2) took place near the E. margin of the rift. The Oslo
Graben (2.2), marked by Permian vulcanicity, may be an early abortive split
and this is certainly true for the North Sea Graben (8.10) with its Jurassic
activity.

A much earlier North Atlantic - Iapetus-played the leading role in Caledon-

Fig. 10.2 "IAPETUS" AT END OF LOWER CAMBRIAN TIMES
(after Owen 1976)

ian tectonics. This started to open some 1000 m.y. ago when a huge block of continental crust, made up of Middle Proterozoic and earlier complexes (2.1), split to form the Eurasian Plate and the northern part of the American Plate. The different palaeomagnetic pole positions of the Lower and Upper Torridonian (2.3) show that this movement was under way in the Upper Proterozoic. These formations, and the Moinian, their broad equivalent in the fold-belt (3.3), were deposited on the margin of the American Plate. As the plates moved apart, however, tectonic events took place on their margins. In parts of the W. side there was the Grenville metamorphism, claimed by some geologists to have affected the "older" Moinian of Scotland and Ireland (3.3 and 3.4).

On the opposite side the 1000 m.y. Sveconorwegian events (2.1 and 2.2) were perhaps contemporaneous with early opening. The much later Cadomian events (5.8) were mainly confined to S.W. Europe, although there was also tectonism in the Midlands kratogen (6.4) and its margins (4.2), and possibly somewhat earlier in Anglesey (4.2).

The Varangian (late Precambrian) glacogenes occur on both sides (2.1, 3.4, 3.5, 3.8 and 5.8) and were probably due to an atmospheric or even cosmic cause rather than to plate movements.

Iapetus seems to have been at its widest in the Lower Cambrian, and the Cambrian Period to have been one of comparative inactivity. Widespread Cambrian quartzites (2.3) were deposited on the N.W. shelf, probably bordering a low, stable, desert land.

Closure was under way by the Upper Cambrian, although the scale of late Cambrian (Sardic) orogeny is doubtful. Sea-floor consumption was earlier in the N.W. and perhaps began earlier in the S.W. of this zone (Connemara, 3.5) before spreading in the Ordovician to the Grampian and Northern Highlands. In parts, at any rate, of the S.E. shelf, however, the Grès Armorican quartzite facies (Lower Ordovician) was deposited from Morocco through Spain to the Welsh Borderland (Ager, 1975, 135). Further N.E. plate collision was under way at this time in Scandinavia and spread progressively to end in the N. in the late Silurian (Gayer, 1973). In Britain the main folding migrated S. during the Lower Palaeozoic. There was, however, a Middle Devonian compression of Lower Devonian, Caledonian molasse further N., again, in the Midland Valley of Scotland (6.2) and in Western Norway (3.8). Space does not permit discussion of the complex and, in detail, debatable, closure of Iapetus (see for example, Dewey, 1971; Mitchell and KcKerrow, 1975; and the summary in a recent volume in this series by Owen, 1976). However, it seems clear that the opposite sides of Iapetus had become welded by the end of the Silurian. The suture most probably passes between the Southern Uplands and the Lake District and continues across Central Ireland (Phillips et al., 1976). Another part lies E. of Trondheim as the Trondheim Nappe (3.7 and 3.8), on faunal grounds, came from the American side.

As Caledonian plate closure was in its late stages in N.W. Europe, a sea was opening further S. in which were deposited the widespread marine Devonian and Lower Carboniferous sediments. These, and the more commonly terrestrial Upper Carboniferous beds with their coals, were compressed in several phases during the Hercynian orogeny. A satisfactory Plate Tectonics model for this orogeny has proved difficult.

The crux of the problem is whether the sea just mentioned was ever an ocean, floored with oceanic crust and separating a northern and southern Europe which had split into two plates. Ager (1975, 137-142), in discussing the views of a number of authors, throws considerable doubt on both the facies and faunal evidence for such a complete separation. There is much less evidence for the presence of ocean floor than for the Caledonides and Alpides; moreover there is no recognisable suture. The problem is also discussed by Owen (1976).

It has been suggested that consumption of oceanic crust took place between Cornubia and Armorica, the evidence put forward being the presence of Devonian and Carboniferous basics and ultrabasics in South West England. Floyd (1972), however, on geochemical grounds, regards the volcanics as continental alkali-basalts. The basics and ultrabasics, too, may be pre-Devonian (5.9) brought up by a thrust-system which splits the Armorican Arc near the N. shore of the English Channel.

Riding (1974) has envisaged oceanic crust further S. between Northern Europe and Africa, with a North Spain "microcontinent" playing a major role in subsequent compression.

In their overall tectonic style the Hercynides differ. Although overfolds and thrusts occur from Spain (5.12-5.14) to South-West England (5.9), the great nappes of the Caledonides and Alpides are not present. Although, too, a Hercynian front can be recognised from the Rhine (5.11) to Ireland (5.10), forming the N. limit of metamorphism and most of the overfolding, powerful folding and faulting persists for several hundred kilometres to the N.

Much remains to be done to find a satisfactory Plate Tectonics model, a task complicated by disconnected blocks and involvement in Alpine tectonics. It is possible that plate separation did no more than thin the continental crust and lead to the development of epicontinental seas and terrestrial basins. Of possible significance is the thinner (23 km) crust on the E. side of the Sillon Houiller fault-zone in the Massif Central (5.2) relative to the W. (30 km). Before this transcurrent fault movement the thinner crust would lie 50 km or so further S.

The Hercynides of Western Europe (apart perhaps from elements now incorporated in the Alps) may, therefore, have been built up on continental crust, admittedly thin in places. It is relevant that there is no consistent migration of tectonism; the Sudetic and Asturian phases, for example, are significant from Iberia (5.12-5.14) to Central Scotland (6.2).

The Variscan and Armorican Arcs would seem to have split round a Cadomian or older block (the northward continuation of the noyau arverne (5.2)) under the Paris Basin (crystallines are known from bores (8.4)), even though this is crossed by the Ardennes (5.11) and possible other Hercynian folds further S. The southward bend of part of the Armorican Arc S. of Brest and La Vendée may be due to curvature round Pre-Hercynian elements known to underlie the Aquitaine Basin (8.4). This part of the Arc continued, prior to the Biscay opening, into Northern Spain where the Cantabrian Arc (5.13) is thought to curve round an ancient massif which may be a westward projection of that under Aquitaine. All this suggests a continental fold-belt, analogous to the Rockies, which contain, among others, the resistant Colorado Plateau.

The Atlantic and Mediterranean plate boundaries remain active and through earthquakes and volcanoes affect man, particularly in the long-inhabited Mediterranean region. Here we may end with perhaps the loudest bang ever heard by man in Europe. This took place about 1500 B.C. when volcanic explosion shattered the Aegean island of Santorini (Thera) and destroyed its Minoan civilisation.

REFERENCES

Ager, D.V. 1975. The geological evolution of Europe. P.G.A., 86, 127-154.
Dewey, J.F. 1971. A model for the Lower Palaeozoic evolution of the southern margin of the early Caledonides of Scotland and Ireland. Scot. Journ. Geol., 7, 219-240.
Dewey, J.F. et al. 1973. Plate tectonics and the evolution of the Alpine system. Bull. Geol. Soc. Am., 84, 3137-80.
Floyd, P.A. 1972. Geochemistry, origin and tectonic environment of the basic and acidic rocks of Cornubia, England. P.G.A., 83, 385-404.
Gayer, R.A. 1973. Caledonian geology of Arctic Norway. Arctic Geol. Mem., 19, 453-468.
Harland, W.B. 1969. Contribution of Spitsbergen to understanding of tectonic evolution of North Atlantic region. Am. Assoc. Petrol. Geol., Memoir 12, 817-851.
Mitchell, A.H.G. & McKerrow, W.S. 1975. Analogous evolution of the Burma orogen and the Scottish Caledonides. Bull. Geol. Soc. Am., 86, 305-315.
Owen, T.R. 1976. The Geological Evolution of the British Isles. Pergamon, Oxford.
Phillips, W.E.A. et al. 1976. A Caledonian plate tectonic model. Journ. Geol. Soc., 132, 579-609.
Riding, R. 1974. Model of the Hercynian Foldbelt. Earth and Planetary Science Letters, 24, 125-135.
van der Voo, R. 1969. Palaeomagnetic evidence for the rotation of the Iberian peninsula. Tectonophysics, 7, 5-56.

INDEX

Aar Massif 162,163
African Plate 153,239
Agde 227,231
Agout Dome 101
Aiguilles Rouges 166
Aix-en-Provence 174
Albertville 168
Algarve 120
Alicante 189
Alpine movements 7,91,132,144,147, 151-190,200,204,214,219,238,242
Alpujarrides 189,190
Alston Block 131
American Plate 239,241
Andorra Massif 183
Anglesey 73,75,76,78,80,219,224,241
Annes Klippe 166
Antrim 87,146
Appin Group 42
Aravis 166
Ardennes 95,112,114
Ardnamurchan 222,224
L'Argentière Window 172
Argyll Group 42
Armorican Arc 98,99,105,108,111,235, 236,242
Arnaboll Nappe 30
Arran 43,128,222
Arve 157
Askrigg Block 131
Assynt 25,30,31,34
Asturian movements 7,108,117,120,121, 242
Atlantic Rift 222,239,242
Avoca 94
Axial Zone 180,182,183

Baerum Caldera 18
Bala Fault 80
Balearic Islands 189
Ballachulish 43
Ballantrae 125
Ballapel Foundation 43,44,46
Ballycastle 43
Baltic Shield 8,13,25,28,56,201,208, 235
Bamble Formation 14
Bandol 177

Banff Nappe 44
Barcelona 186
Barents Sea 56,61,68
Barr-Andelot Granite 103
Basel 157
Bath Axis 142,213,214
Bay of Biscay 118,180,185,201
Belfast 86
Belgian Coalfield 95
Belledonne Massif 168,170
Ben Nevis 40,48
Bergen 57,65
Berne 157
Bernesga Valley 118
Betic Zone 189
Bilbao 185
Biscayan Zone 185
Bonn 105
Bordeaux 203
Bornholm 13
Borrowdale Volcanic Series 81,83,84
Bort 231
Boulonnais-Artois Anticline 201,207
Bovey Tracey 111
Brabançon 6
Brabant 94
Bragança 120
Bray Anticline 207
Bray Group 91,92
Brèche Nappe 159
Brecon Beacons 139
Bretonnic movements 7,108,117,120
Briançonnais Zone 168,172
Bridgwater 110
Brioverian 101
Bristol 144
Bristol Channel 110,139,143,219,239
Brittany 105,107
Buchan district 46
Builth 76
Burgundian Gate 155

Cader Idris 75
Cadomian movements 6,107,109,110,120, 235,241
Cairngorm Mountains 46
Calatayud-Teruel Graben 186
Caledonian Front 8,13,14,18,20,25,28, 30,31,56,65,236

Caledonian intrusions 35,43,48,63,80, 83,88,89,131,136
Caledonian movements 6,80,85,88,92, 94,115,125,130,147,236,241
Caledonian suture 236,241
Callander 43
Cambrian 18,20,25,30,31,43,57,59,61, 63,73,75,81,91,94,101,112,114,117, 118,134,142
Cannes 177
Canonbie 89
Cantabrian Arc 117,242
Cantabrian Zone 115
Cantal 229
Cape Wrath 20
Carboniferous 46,76,83,87,92,105,110, 112,114,117,118,127,128,130,134,140, 142,146,165,168,172,183,206,208,216
Carcassonne Gap 201
Cardiff 141,142
Cardigan Bay 76,81,219
Careg Cennen Line 73,78,139
Carlisle 83
Carmel Head Thrust 80
Carn Chuinneag-Inchbae Granite 35,37 48
Carrauntoohil 111
Les Causses 98,101
Celtic Sea 219
Central Graben 218,238
Central Highland Psammitic Group 42
Central Iberian Zone 118
La Cerbère 183
Chain of Puys 229
Chambéry 155,156,166,168
Chamonix 166,168
Channel Islands 105,107,108
Charleroi 114
Charnian 134,136
Chepstow 139
Cherbourg 107,108
Cheviot 128
Church Stretton 73,78,132,139,236
Cimmerian movements 7,210
Clare 147
Clare Island 40,51
Clermont-Ferrand 101,227
Clew Bay 40,50
Clyde Plateau Lavas 125
Connemara 48,50,73,241
Córdoba 120
Cotentin 107
Craven Faults 128,132
Creeslough Succession 43,46
Cretaceous 13,44,63,95,160,168,170, 174,177,182,183,185,187,195,203,206,
208,213,216
Croagh Patrick 51
Cromarty Firth 39,40
Culoz 157

Dala sandstone 12
Dalradian 42,48,51,54,145,227
Dalslandian 10,13
Danian 13
Dauphiné 168,194,238
Dauphinois Zone 170
Dent Blanche Nappe 165
Devoluy 170
Devonian 100,105,108,110,112,114,117, 118,206
Diablerets Nappe 162
Die 168,170
Digne 170,174,177
Dijon 101,193,201,204
Dinant Synclinorium 112,114
Dingle 111,112,144,145
Dingwall 39
Dinorwic 80
Dodman 110
Donegal Granite 43,48
Dornoch Firth 39
Douro Basin 120
Downtonian 18
Dublin 91,92,144,145
Dungarvan 111
Durance Valley 170,172,174
Durness Limestone 25

Eakring 212
East Pennine Coalfield 136,212
Les Ecrins 170
Edinburgh 128
Egersund-Ogna anorthosite 17
Eifel 231,233
Ekofisk Oil Field 218
Embrunais Nappes 153,172
Engadine Window 165
English Channel 218
Eocambrian 13
Eocene 25,183,195,197,199,204,210, 213,218
"Eo-Europa" 3
Etive Complex 48
Eurasian Plate 239,241

Faeroes-Shetland Basin 219
Faille du Midi 112,114,206
Fannich 37
Fen 17
Finnmark 12,13,56,57,61,67
Finse district 61

Firth of Clyde 43
Fishguard 75
Folded Jura 155
Franco-Swiss Plain 153,157
Fyn-Ringkøben High 218

Gaissa Nappe 57,59,67
Galway 50,88
Gap 168,170
Gavernie Nappe 183
Geneva 153
Gérardmer 103
Gergovie 229
Gerona 186
Gibraltar Complex 190
Girvan 125
Glarus 160
Glasgow 127
Glen Coe 48
Glenelg 37
Glenfinnan Division 35,37
Glen Roy 46
Gloucester 132
Gothard Massif 163
Gothenberg 10
Gothian 10,12,13
Gotland 13
Gower 141
Grampian Highlands 28,34,40
Granada 187
Grande Faille des Vosges 103
Grand Paradis Massif 168
Grand St. Bernard Nappe 165
Grasse 174,177
Great Glen Fault 28,34,40,50,54
Great Whin Sill 130
Grenoble 170
Grenollers Graben 186
Grenville metamorphism 241
Grès Armorican 241
Grisons Basin 165
Grondu Thrust 210
Groningen Gas Field 208,216

Halokinesis 185,200,204,210,218
Hampshire 213
Haperanda 12
Harlech Dome 75,80
Harris 22
Heidelburg 103,199
Helmsdale Fault 39
Helvetic Nappes 153
Helvetides 160
Hercynian Front 28,110,111,112,123,
   142,143,144,147,236,238,242

Hercynian movements 7,25,39,50,53,78,
   80,85,89,94,95,100,108,110,114,117,
   120,121,123,127,128,131,136,141,142,
   147,182,238,241,242
Hereford 132
Highland Boundary Fracture-Zone 40,
   44,48,73,123,127,144,146
Hohe Tauern Window 165
Huy 114

Iapetus 239,241
Iberian Pyrite Belt 121
Icartian 6,107,235
Iltay Nappe 43,44,46
Inishtrahull 22,31
Inner Hebrides 20
Innimore Fault 40
Insubric Line 166,239
Interlaken 163
Inverian 6,23,235
Irish Sea 86,88
Irish Sea Geanticline 219
Isère 168
Islay 24,31
Isle of Man 81,83,85,86,219
Isle of Wight 213,215

Jaca 183
Jamtland Supergroup 59
Johnstone Thrust 142
Jotnian 10,12
Jotunheim 61,65
Jurassic 13,39,63,76,98,101,134,160
   170,172,174,190,193,197,203,206,213,
   216

Kaiserstuhl 231
Kandersteg 160,162,163
Karelian 10,57,235
Katarchean 10,22,235
Killarney 111
Kilmacrenan Succession 43
Kingscourt Outlier 87,89,91
Kintail 37
Kintyre 44
Kirkenes 12
Kiruna 11,12
Kishorn Nappe 31
Kola Peninsula 5
Kolsä 18
Kolvik Thrust 59
Kongsberg Formation 14,15
Kristiansand 14,15

Laacher See 233

Lake Mjösen   13,56,61,63,65
Lake Vänern   10,12
Laxfordian   6,23,235
Leadhills   89
Leannan Fault   50
Leinster Batholith   94,147
Leny Limestone   43
Léon (France)   107
León-West Asturias Zone   115
Lewisian   20,22,30,34,35,46,54,56,236
Liassic   25,44,83,140,144,203
Liège   114
La Limagne   101,229
Limerick   145,146
Lizard   110
Llanberis Fault   81
Llanes   118
Lleyn Peninsula   75,78,80,219
Loch Alsh   20
Loch Awe   43
Loch Don Anticline   227
Loch Eil Division   35,37,42
Loch Eriboll   30
Loch Linnhe   34
Loch Lomond   43
Loch Maree   22,24,25,31,39
Loch Ness   34
Loch Ryan   89,91
Loch Skerolls Thrust   31
Loch Tay Fault   50
Lofoten Islands   67
London   212
Longmyndian   75,134
Lough Derg   46
Lough Neagh   88
Luchon   182
Lugnaquillia   91
Luna Valley   118
Lundy Island   111,219,224

Macon   193
Mainz   105,197
Maladetta Massif   183
Malaga   189
Malvern Axis   142,143,144
Malvern Hills   132,136,137
Maritime Alps   174
Market Weighton Axis   213
Marseilles   174
Matterhorn   165
Maures Massif   174,177
Mayhill Anticline   137
Median Pre-Alps Nappe   159
Mediterranean Volcanic Belt   177,222, 227,242
Médoc-Montauban High   204

Menai Straits   81
Mendip Axis   142,143,213,214
Mercantour Massif (Argentera)   170,174
Merrick   86
"Meso-Europa"   3
Meuse   94
Midlands Kratogen   28,235,236,241
Milford Haven   139
Minch   22
Miocene   155,157,174,186,187,195,197, 199,204,207
Mischabel   165
Mochras Bore   76
Moffat   86
Moine Nappe   30,31
Moine Thrust   20,30,31,37
Moinian   31,34,35,39,42,46,48,54,241
Monadliath Mountains   46
Monian   75,78,80
Mont Blanc   151,166,168
Mont Dolin   165,166
Mont-Dore   98,229
Montélimar   193
Monte Rosa   165
Mont Peyroux Nappe   101
Montsech   183,185
Monts Salève   157
Morar Division   35,37
Morcles Nappe   160
Morvan   98,100
Mount Sorrel   136
Mulhacen   187
Mulhouse   197,231
Mull   222,224
Mythen   163

Namur Synclinorium   112,114,207
Nappe des Brèches   168
Nappe des Gypes   168,172
Neath Disturbance   137,141,142
"Neo-Europa"   3
Neogene   81,200
Nesodden   18
Neuchâtel   156,157
Nice   174,177
Niesen Nappe   159
Norbotten   10
Normandy   105
Norrland   11
Northern External Zone   180,182
Northern Highlands   28,34
North Pyrenean Fault   182
North Rona   22
North Sea Oil and Gas Field   216

Oceanic Crust   239,242

INDEX

Ochil Hills 123
Offerdal Nappe 59
Old Red Sandstone 35,39,43,44,48,50,
  54,56,63,76,83,87,89,92,125,127,130,
  131,134,139,142,145,216
Oligocene 101,111,146,157,170,174,183,
  195,197,206,213
Omagh 43
Oporto 120
Ordovician 18,25,43,51,53,57,59,63,
  73,75,81,87,92,94,117,118,125,145,
  183
Orkney Islands 28,35,39
Orleans 207
Oslo 12,18
Oslo Graben 14,18,239
Ossa-Morena Zone 118,120
Ostersund district 59
Outer Hebrides 20,22,24
Oviedo 117,118
Oxford 213
Ox Mountains 46

Palaeocene 206,210,218
"Palaeo-Europa" 3
Palaeogene 199
Pamplona 183
Paris 206,208
Pau 203
Peak District 136
Pebidean 75
Peel Horst 210
Pelitic and Quartzitic Transition
  Group 42
Pelvoux Massif 170
Pennides 155,165
Pennine Fault 81,85,86,128,131
Pennines 128,130,132
Pennine Thrust 163
Penteverian movements 107,235
Pentland Hills 125,127,128
Permian 14,18,44,83,87,98,103,125,
  131,134,146,160,162,170,177,203,206,
  208,213
Petsamo 12
Phanerozoic sequence 4,5
Pico d'Aneto 177
Picos de Europa 115
Piémont Zone 153,159,168
Plateau Jura 155
Plate Tectonics 239,241
Pliocene 213,218
Plynlymmon Anticlinorium 80
Poitiers 101,201
Pontypridd Anticline 141
Porcupine Bank 34

Porsangerfjord 57,59,67,68
Pre-Alps 153,157
Prebetic Zone 189
Precambrian, Late, correlations 2
Pregothian 6,10,12,235
Prekarelian 10
Provence 174,177,238
Puigcerda 180,183
Purbeck Anticline 215
Le Puy 231
Puy de Dôme 229

Quarff Succession 56
Quaternary sediments 195,213
Quaternary volcanics 98,105,195,229,
  231

Research, Early 1
Rhaetic 13,44,134,140,193,208,213
Rhine Graben System 95,103,105,155,
  195,201,231,238
Rhône 156,160,163,174,193,194,227,
  231,238
Rhum 222
Rif Mountains 190
Rio Gallego 185
Ritec Fault 139,142
Riviera 151,238
Rockall 20,219,224
Rocroi 115
Rogaland 14,15,17,20
Rondaides 190
Root-Zone (Pennides) 165,166
Rosslare Complex 91,92,94,236

Saalian 7
St. Brieuc 107
St. Etienne 98
St. George's Land 140
St. Kilda 224
St. Michel 172
St. Moritz 166
Salrock Fault 53
Sambre-et-Meuse Massif 112
San Sebastian 185
Sanquhar 89
Santander 185
Santorini 243
Saône 193,195
Sardic movements 44,48,241
Sarv Nappe 59
Satellite Massifs Zone 180,182
Saxonian movements 7,204,210,238
Scafell Pikes 81
Scania 13
Schonau 105

Schoonbeck Oilfield 210
Scourian 6,22,235
Série de Fully 166
Serre Inlier 103
Sgagerrak 14
Sguir Beag Slide 37
Shelve 75
Shetland Islands 28,54
Siccar Point 86,89
Siebengebirge 231
Sierra Nevada 187,189
Sillon Alpin 170
Sillon Houiller 99,207,231,242
Silurian 18,51,53,59,63,68,73,76,83, 85,87,92,95,120,125,134,136,137,139, 142,145,151,177,182,213
Simme Nappe 159
Simplon-Ticino Nappes 165,166
Sisteron 170
Skiddaw Slates 81,83,84
Skye 24,31,219,222
Slieve Gullion 224
Slioch 24
Småland 12
Snowdon 73,75,78
Sogndal 17
Somport Pass 180,182
Soria 187
Southern External Zone 183
Southern Highland Group 42
Southern Uplands Fault 88,91,127
South Portuguese Zone 118,120,121
South Pyrenean Fault 183
South-West England granite 108,110
Sparagmitian 13,63
Start Point 109
Stavanger 17
Stavolot 114
Stephanian 100,108,118,120,170,177, 203
Stoer Group 24
Storefjeld 17
Stornoway 24
Strabane 43
Strasbourg 197
Strathmore Syncline 127
Subbetic Zone 189
Sub-Briançonnais Zone 166,168,172
Sub-Jotnian 10,12
Sub-Pyrenaic Sierras Zone 183
Sudetic movements 7,108,143,242
Sulens Klippe 166
Svalbardian 6
Svecofennian 10,11,235
Sveconorwegian 6,10,13,17,235,241
Swabian Plateau 231

Swansea Valley Disturbance 141,142
Swedish Border Mountains 59,61,65

Table Jura 155
Taconic movements 6,68
Tanneron Massif 177
Tarragona 185
Tarskavaig Nappes 31
Tavetscher Swischenmassiv 163
Telemark Formation 14,15
Tellenes 17
Tertiary intrusions 212,219,222,224, 225,227
Tertiary movements 39,50,53,68,86, 101,108,111,118,120,121,128,132,136
Tertiary volcanics 98,105,174,186, 187,195,222,229,231
Tethys 151,239
Thun, Lake of 157,160
Torridonian 20,24,30,31,241
Toulouse 203
Towy Anticline 76
Triassic 13,24,25,39,44,76,83,86,87, 98,103,125,134,140,155,156,170,172, 174,182,185,187,189,193,203,206,208, 213,216,219
Trondheim 61,63,65,241
Trysil sandstone 12
Turin 153,172,239
Tyrone Inlier 145

Ullswater 81,84,85
Ultradauphinois Zone 170
Ultrahelvetic Nappes 159,160
Unst 56
Uriconian 75,78,134,136

Valaisian Zone 168,172
Valdres Sparagmite 65
Valence 195
Valensole Basin 174
Vallorbe-Pontarlier Fault 157
Vanoise Massif 168
Varangerfjord 12,13
Varangian glacogene 43,51,57,59,63, 107
Variscan Arc 98,99,101,242
Vélay 231
La Vendée 105
Ventoux Line 170
Vercors Massif 168
Veynes 170
Vigesa 17
Viking Graben 218,238
Visingso Formation 13
Vocontian Trench 168,170

Vogelsberg 231

Wallensee 160,162
Walls Boundary Fault 54,56
Wash 212
Weald Anticline 212,214
Weardale Granite 131
Wensleydale Granite 131
Westerwald 231
West Minch Fault 219

West Shetland Shelf 22,219
Weymouth 214,215
Wildhorn Nappe 162
Window of Ord 31
Window of Theux 114
Woolhope Anticline 137

Zaragoza 199
Zurich 155